Science Past—
Science Future

Essays on Science by Isaac Asimov

From *The Magazine of Fantasy and Science Fiction:*
FACT AND FANCY
VIEW FROM A HEIGHT
ADDING A DIMENSION
OF TIME AND SPACE AND OTHER THINGS
FROM EARTH TO HEAVEN
SCIENCE, NUMBERS, AND I
THE SOLAR SYSTEM AND BACK
THE STARS IN THEIR COURSES
THE LEFT HAND OF THE ELECTRON
ASIMOV ON ASTRONOMY
ASIMOV ON CHEMISTRY
OF MATTERS GREAT AND SMALL

From Other Sources:
ONLY A TRILLION
IS ANYONE THERE?
TODAY AND TOMORROW AND—
SCIENCE PAST—SCIENCE FUTURE

Science Past—
Science Future

ISAAC ASIMOV

DOUBLEDAY & COMPANY, INC.

Garden City, New York 1975

Library of Congress Cataloging in Publication Data
Asimov, Isaac, 1920–
    Science past, science future.
    1. Science—History.   2. Technology—History.
3. Science—Social aspects.   4. Technology—Social
aspects.   I. Title.
Q125.A77   508′.1
ISBN 0-385-09923-1
Library of Congress Catalog Card Number 74–25092

To my wife, the novelist,
and her book, *The Second Experiment*

Contents

Two —

Three *Science Future*

Introduction

On May 30, 1974, I set sail on the *France* for Great Britain. It was my first trip to Europe since I, as a three-year-old, had left that continent with my family to take up permanent residence in the United States.

I approached the whole thing with considerable dread because I don't travel, as a usual thing. Still, two years of negotiation had extorted from me a pledge to go to London to give talks, to sign books, and to do various other minor-celebrity things—and I was committed.

Fortunately I had a good time, temporarily gained ten pounds, and loved everything about the land and its people. What I loved most was the unfailing politeness of the British people.

I live in New York City, you know, and our way of life here is adapted to the necessities of huge crowds surging this way and that. The easiest way of handling the difficulties imposed on us is, I suppose, to pretend that no one is there, so the average New Yorker ignores everyone. This is not out of rudeness or indifference, but out of the necessity for self-preservation. I must admit, however, that it *looks* like rudeness or indifference. Therefore, when I encounter a civil verbal interaction between strangers, it seems rather charming to me.

I was most aware of this in Stratford-on-Avon, which, as a writer of a two-volume book on Shakespeare, I *had* to visit. I'm told that the crowds in Stratford are almost entirely American tourists, with only just enough natives to man the curio shops, but everything seemed British to me.

I had tickets for a performance of *King John*, and had a cafeteria meal beforehand (you know, sausage roll, steak and kidney pie, treacle tart—all the items everyone warned me would corrode non-British intestines, but which merely helped shovel the pounds on me).

I was at the rear of the line when I noticed that at the head of the line was someone who had come to within six feet of the cashier and was left standing there, while the person behind the counter went off to see if she could find a helping of roly-poly pudding (or whatever) for him. He waited with the greatest of composure. There was no sign of impatience about him and I noted with astonishment that he did not rap on the glass top of the counter with his umbrella; not even once.

Behind him a long line of British men and women waited with equal composure, while the cashier, with nothing to do, gazed thoughtfully into the middle distance. In New York, of course, the line would have surged past the man at the head as though he were not there—acknowledging his presence, if at all, only by a variety of scowls, each designed to fit the personality of the scowler.

Finally, the second man in the line stirred. Instead, however, of taking decisive action, he cleared his throat and said to the first man, "I hope you won't mind if I overtake you."

"Not at all," said the person addressed, with grave courtesy. "Pray do."

And the line moved at last, each stepping past with a kind of grave apology.

After the play was over and we were leaving the theater, my wife stopped short and said, "Oh, dear, I think I've left my belt in my seat." (She hadn't, as it turned out, but had left it back in our hotel room. However, neither of us knew this at the time.)

As a New Yorker, I was galvanized into action by the instinctive knowledge that although the belt might have been unaccountably overlooked by the predators in the theater, this could not be so for long, and the only chance I had of recovering the belt was to get back to the seat in the very minimum of time.

"Wait here," I said, urgently, "and don't move a step. I'll be right back." And with that I plunged into the stream of humanity emerging from the theater.

There is virtually nothing anyone can teach a New Yorker about broken-field running, and I weaved my way against the tide with consummate skill. Still, the best of us have our bad moments, and when I had nearly reached the seat, I misjudged a maneuver that should have twisted me safely about a solidly built woman with the usual quarter inch to spare. Instead, I hit her in the shoulder, rather hard, and sent her spinning up the aisle.

I stopped in horror. In New York, no bypasser notices such a small thing, but here, it seemed, everyone stopped to look. In New York, furthermore, I would certainly have become the target of some animadversions upon my appearance and ancestry on the part of the aggrieved woman, and might even, perhaps, have received a shrewd buffet or two from a purse or umbrella.

And as I stood there, frozen, the woman recovered, staggered toward me, and said, in the quietest possible way, "I do beg your pardon, sir."

To which I could only reply, "Not at all, madam. Pray do not concern yourself about it."

We parted friends.

Somehow I feel I relive this event constantly, for my life is one continued assault on my Gentle Readers.

Let us not even speak of the total number of my published books which, at the moment of writing, is over 160. If we concentrate instead on my books of science essays only, you can see by the list at the front of the book that this is my sixteenth.

I suppose that sixteen collections of science essays is enough to send all my readers spinning up the aisles. And yet I get no sign of this from the letters I receive. Rather, I tend to get letters explaining that the letter writer has read only some half dozen or so of these books and is feeling a little apologetic at not having gotten to the rest of them yet.

What can I reply but, "Not at all, madam (or sir). Pray do not concern yourself about it."

But, of course, if you *should* buy all my essay books (or all my books, without qualification, for that matter) and read them all, I would have no serious objection.

And no nonserious objection either, if it comes to that.

ISAAC ASIMOV

August 1974

NOTE ADDED IN PROOF:

These more than forty articles were originally written at different times for different audiences. There is occasionally some small overlapping among them. For that, I ask your pardon, Gentle Reader.

ONE

Science Past

A · Technology

1 · Technology and the Rise of Man

If you think we are living in an age of technology, you are perfectly right. If you think this is a phenomenon of recent years or of recent centuries, you are entirely wrong.

Technology is the knack of doing things by means of objects that are not part of your own body. If you try to crack a nut with your teeth, you are being natural. If you place the nut on a rock and hit it with another rock, you are being technological.

If we go far enough back in history to reach a time when man did not use technology; when he did not form or shape tools out of rock, bone, wood, or reeds; then we find we must go back beyond man altogether. Our earliest traces of the most primitive manlike creatures, who lived in East Africa 3½ million years ago and who had brains scarcely larger than that of a modern gorilla, are already accompanied by shaped rocks, clearly designed to cut and chip.

What drove man to technology? Surely it must have been the realization, however dim, that by adapting the material in the world about him to his purposes, he could make himself more comfortable. By using the materials at hand, he could kill an animal more easily and be more sure of having food, or he could build some sort of shelter and be better protected against a storm.

If there is but the brainpower and insight to see a way of doing something desirable more easily by means of something outside yourself, it is inevitable that that way will be taken.

The first major step up the ladder of technology came with the reasoned use of fire. This is more than the discovery of fire itself; every animal discovers fire when he has to dash out of the forest that has been set ablaze by lightning. What was needed was more than mere discovery that the thing existed.

There must have come times when men (or children, perhaps), observing a dying remnant of fire consuming some twigs, dared play with it and feed it more twigs. Then it occurred to someone that a *tame* fire, kept in a person's shelter, would give light and warmth at night. It was when that was put into action that the single most crucial step in man's history was taken.

We don't know when this happened, or how long ago, or what the circumstances were, or how many times the discovery was made and remade, or how fast the technic spread.

Eventually, though, it spread to all of mankind. No group of men has been found anywhere on Earth in historic times who lacked the use of fire. It is this use of fire that represents the first absolute distinction between man and all other species of living creatures. There are animals that communicate effectively, even if not by speech; there are animals that use tools in primitive fashion; but no animal other than man has ever made even the faintest beginning toward the taming of fire.

Fire had as its immediate effect, no doubt, the fulfillment of what must have been its immediate purpose—the warding off of the darkness and chill of night. It meant that human beings could be more efficiently active at night, while other large (and possibly dangerous) animals, which feared fire, had to stay away. Man suddenly gained a large advantage in his struggle against other predators.

The consequences of an important technological advance cannot possibly be confined to immediate effects. The influence of fire spread out.

Man was, in his origins, essentially a tropical animal, as almost all other primates are. With fire, however, the colder nights at greater distances from the equator and at greater heights above sea level became bearable. The cold desolation of the northern winters could be endured once man learned to wear the skins of other animals to make up for his own hairlessness, and once he had fire as his ally.

Fire thus meant an outward extension of man's living range, and this has been the effect of every advance in technology since. Man is not confined to those environments to which his naked, unaided body is adapted. Man need not require the age-slow rate of evolutionary change to fit him to other, harsher environments. By technology, he controls the environment and fits *it* to *his* needs.

Nor can all the broadening effects of a crucial technological advance be logically foreseen. There are always completely unexpected and perhaps unpredictable consequences. Where fire was concerned, the discovery was made, undoubtedly by accident at first, that food that was heated by fire became easier to chew and developed new flavors that men found pleasant. Gradually, the use of cooked food became universal.

This was not just a matter of fashion or of needless artificiality. There are some foods that can be eaten when cooked but that are too hard or coarse to serve man in the uncooked state. Cooking softens and, to an extent, predigests food, and these are sometimes essential. Consider how many people are supported by cooked rice, for instance, and imagine eating uncooked grains of that cereal.

Then, too, cooking destroys bacteria and other parasites in food. The use of cooked food cut down on disease and worm infestation in primitive man—and it meant an increase in strength and a lengthening of life.

Fire, by increasing the food supply in this way, made it not only possible for man to increase his range but also to increase his numbers. The same quantity of land, the same supply of plants and animals, could, with fire, support more human beings than it could without fire. And ever since, further advances in technology have further increased the number of human beings the Earth could support at one time.

The effects of a crucial advance in technology continue to ripple outward and sometimes do not make themselves evident for surprisingly long periods of time.

Early man may occasionally have found small nuggets of copper, silver, or gold and been fascinated by the shine, their ability to be flattened and deformed into interesting shapes without breaking, the ease with which they could be polished to a gloss. It would have been natural to use the nuggets as ornaments.

Eventually, about 3500 B.C., it was somehow found that certain rocky minerals, when mixed with charred wood, could yield large quantities of certain metals—provided the mixture was heated strongly, by the action of fire. The use of metals became possible only because man had fire at his disposal, so that new technological advances build on old ones. And metals came into use perhaps tens of thousands of years after fire was first tamed, so that consequences are sometimes delayed.

Of course, the use of fire involved dangers. Fire might have been tamed, but it was always ready to go wild again. A campfire might accidentally destroy the shelter within which it burned—or an entire forest, for that matter. Men might die in flames or even deliberately use fire as a weapon to kill other men.

There is no way of preventing such tragedy if man is careless or malicious; but the point here is that it is not the technic that produces the danger but the man who controls it. Used efficiently, carefully, and humanely, there is no reason to suppose that fire would be destructive.

But are there not ways in which fire, however efficiently, carefully, and humanely used, would produce undesirable side effects? Surely, it produces a smoke and reek that fouls the air, smudges all with which it comes in contact, and irritates the throat and lungs. Many a fire user might (we can imagine) sigh for the clean air of a tent or cave not polluted by smoke.

The alternatives are to abandon the use of fire, to use the fire and endure the pollution, or to think of a way of having the fire without pollution. The last alternative is surely the most reasonable, and it was, indeed, the one taken. Some chimneylike device was developed that would allow the smoke to escape out of the shelter and into the open air, where it was more effectively diluted.

This is an example of the manner in which the problems caused by technology are solved by additional technology. This, in turn, may produce new problems—thus, man's fires have now increased in quantity and intensity to the point where even the entire atmosphere threatens to be fouled by them. What do we do now? Abandon, endure, or move forward with new ingenuity? All history shows that only the third alternative—new ingenuity and more advanced technology—is tenable.

Technological advance brings with it a less material discomfort in the form of a growing dependency of man upon the technology. If it is fire that keeps you from freezing in winter, then you must run the risk of freezing if the fire should go out accidentally. If you had remained without fire and stayed in the tropics where you belonged, you would never run the danger of freezing.

But the answer to dependence, also, is further advance. At some time, man learned how to start a fire deliberately where no fire had previously existed, by sparks or friction. When that new technic was developed, the fear of having a fire go out vanished.

If we trace the consequences of the use of fire, and of other advances in technology that the use of fire made possible, and the still other advances that *these* advances made possible, it becomes tempting to argue that all changes in the human condition depended, one way or another, on some advance in technology. Even changes that seem, at first glance, to have nothing to do with technology can be traced back to some technological factor that made the change possible, desirable, or both.

Let us pass on to the greatest technological advance after the taming of fire—the development of the technics of agriculture. This came about 10,000 B.C., somewhere in what we now call the Middle East.

In a food-gathering economy, individuals must have come across certain wild grasses, with seeds that were tasty and nourishing once roasted. It must have occurred to someone that if some of the seeds were planted instead of being eaten, and if the seedlings were cared for, there would be a sure supply of food a year later.

Who thought of this, how the thought arrived, how often the attempt was made before it succeeded, how people learned over what space of time and with what difficulty how best to care for the growing plants, how to hoe and weed and water and fertilize, how to process the final grain, we don't know—but eventually small farming communities were founded.

(The taming of plants was analogous to the taming of animals—the deliberate collection of herds that would be cared for, fed, and encouraged to breed in order that there might be a ready meat supply brought to the table, without the dangers and uncertainties of hunting. Herdsmen, as well as farmers, came into being.)

Consider the consequences of agriculture. For one thing, if a tract of land was given over to the careful cultivation of plants suitable for food, many more people could be supported than if that land were left in its natural state. Population therefore increased unusually rapidly in those areas that were devoted to agriculture, and the human race underwent its first population explosion.

But as the population grew, the stakes grew higher in case of catastrophe. If the harvest failed, the number of people crowded onto the land in anticipation of plentiful food could not be supported in any alternative fashion. Starvation would be more widespread and serious than in the days of thinner population supported by hunting and food gathering.

Abandon? Endure? Advance? It was advance, of course. Since the most likely cause of poor harvest was the failure of rain, farming had to be done where water was available, even in the absence of rain—that is, in river valleys. What's more, a network of irrigation canals had to be built and maintained in order to make certain that river water would reach all parts of the agricultural territory. Dikes would have to be built to contain flood waters when the river's water level grew too high.

To build and maintain the canals and dikes required the co-operation of many people. This meant that agricultural cultures could not be organized on the basis of families, as food-gathering cultures could be. There had to be a larger and more inclusive political organization, and so the first city-states developed in the farming areas along the banks of the Nile, the Euphrates, and the Indus rivers. They developed not because men had developed some sort of abstract political notion, but as a direct consequence of the advance of technology.

What's more, the development of agriculture meant that, for the first time, a group of people could produce more food than was required for their own needs. This food surplus could support individuals who did not directly contribute to increasing the food supply but who could provide goods or services of other sorts that the farmers valued and would trade their food for.

It meant that mankind could specialize. It meant that artisans could exist, and merchants, and poets, and artists, and priests. The life of the intellect began, not because mankind somehow grew more intelligent or because someone had an abstract idea— but only because technology had finally supplied enough food to allow this basically parasitic type of activity to exist.

Naturally, the supply of food accumulating in agricultural regions attracted the "barbarians" outside, who were still gathering, hunting, or, sometimes, herding food. Barbarian raids, which led to the slaying of farmers and the rifling of granaries, made it necessary for the farmers to band together for mutual protection. Houses were huddled together and surrounded by walls.

Thus began the city, and with it "civilization" (from the Latin word for "city"). Within the city, wealth accumulated, and those who were not farmers remained permanently. Organized warfare began, with the advantage often on the side of the more active and mobile barbarians, who periodically conquered an agricultural territory and then settled down to become farmers in their turn and suffer barbarian depredation.

It was not till the city specialists developed war weapons that depended so massively on advanced technology that the barbarians could not duplicate them, that the pendulum swung massively toward the side of the city dweller. It was gunpowder, developed in Europe in the fourteenth century, that proved the tide turner.

It is no accident that it was in the thirteenth century (the one *before* gunpowder) that the last great wave of barbarians—the Mongols of Jenghiz Khan—devastated the civilized world, West and East. It is also no accident that in the fifteenth century (the one *after* gunpowder), Europeans and their guns began the process that was to place all the world in their power.

But back to farming . . .

The production of excess food that led to specialization in all forms of artisanry also led to trade as the products of one culture were exchanged for the products of another. The desire to trade was made effective by the development of the technologies of road building and of ship building. Through trade neighboring peoples grew to know one another more intimately and could co-operate (or quarrel) more effectively.

As the means of communication and transportation grew with steady advances in technology, larger political units (empires) became feasible and were established. In general, empires grew larger, tending to reach an extent that the technology permitted. The Persian Empire was cemented by its couriers, who rode tirelessly over roads at the behest of the Great King. "Neither snow, nor rain, nor heat, nor gloom of night" (said Herodotus) "stays these couriers from the swift completion of their appointed rounds." But the technological basis for holding so large a realm as the Persian Empire together was too slender, and under an unenergetic monarch, when it had to face the energetic Alexander the Great, it fell apart.

The largest and most successful of the ancient empires in the West was the Roman Empire, and what made it so successful? Was it the dedication of its citizens, the excellence of its laws, the wisdom of its rulers, the valor of its armies, the cleverness of its generals? No doubt, all contributed (though it failed often enough in each of these categories), but the Empire could not have endured, though it had all of these, if it had not also developed the Roman roads along which the legions could march from end to end of the Empire, whenever and wherever their services were required. "All roads lead to Rome," the proverb says.

Even when the advance of technology does not lead directly to some advance, it can set up conditions that make an advance necessary. As city-states grew toward empire, thanks to the advance in the technology of transportation and communication, the financial conduct of the growing unit became more complicated. It became more and more difficult to organize and control the statistics of taxation. In order to tally the items demanded and compare it with those received, in order to keep track of reserves and expenditures, some agreed-upon marks had to be made to assist the memory, and in the end (shortly before 3000 B.C.), by natural stages, writing was developed.

Before writing became efficient, a variety of technics had to be developed—a stylus to punch marks on soft clay, and the clay being then baked to permanence; a brush and ink to make marks on a paperlike surface manufactured from river reeds.

The development of writing, which arose out of a need produced by the advance of technology, itself hastened further advances. Writing meant the more or less permanent storing of knowledge. No longer would vital knowledge die at once with the memory that had held it. Literary products multiplied; law codes were written down for permanent records, as were the chronicles of the time; knowledge accumulated.

With recorded details of thoughts and deeds available in the cities, it became easier to adopt advances already adopted elsewhere, or to use earlier advances as springboards for still newer advances.

In short, it is the lesson of history that the rate of advance of technology has constantly increased, as one group of discoveries serves as a basis for later, more extensive ones. Technology, in fact, is the one aspect of human endeavor that has never (so far, at least) retreated. Even in so-called Dark Ages, when literature and theoretical speculation fade and the amenities of society decline, technology continues to advance, though perhaps at a slower rate. The centuries of decline after the fall of the Roman empire saw the introduction of horseshoes and horse collars; the moldboard plow; Gothic architecture; the magnetic compass; gunpowder—each of which had a powerful effect on society, and the last two of which clearly enabled Europe to conquer the world.

Does farming bring nothing but advantages? Of course not. No technological advance can fail to have its undesirable side effects.

Man's effort to concentrate on those particular plants useful to himself offered immense riches to those other animals who fed on those plants, so that all kinds of "vermin," from rats to locusts, multiplied, sometimes to the point where their depredations left starvation for man in its wake. The multiplication of man's own numbers led to the easier contagion of disease, and great plagues now and then afflicted mankind.

The connection between these ills and agriculture is not really obvious. What was much more clear was that farming tied man to the soil. He was no longer free to roam and hunt. He had to stay within reach of his unmoving plants, to care for them and to defend them. He had to work backbreakingly hard in every aspect of the art. And because it was the excess food he produced that made it possible for courts and armies and cities to exist, the ruling groups in the ancient empires saw to it that farmers worked as long and as hard as possible—to the point where slavery became common.

All through postagricultural times, there had to be a longing for a return to the "simple life." Why else do so many mythologies include tales of a Golden Age of the past when food was simply gathered without effort? The best-known tale of this sort is, of course, that of the Garden of Eden, from which Adam was driven out with the curse, "In the sweat of thy face shalt thou eat bread."

Yet though men might idealize the food-gathering society out of all recognition and be distressed over the curse of farm work, there was no going back. The advance of technology had meant an increase in population and the creation of new comforts. To undertake a retreat in technology would mean a sharp decrease of population (by mass starvation or mass killing) and the loss of those comforts.

Faced with that cold fact, only scattered individuals here and there have ever retreated to the "simple life." No matter how much they urged it on others, the population generally could not follow; they literally could not. No farming community in history, anywhere, at any time, has voluntarily and en masse abandoned farming and resumed food gathering. It is not possible to make such a change.

(And this holds true for *every* important technological advance. Any retreat to a previous level must mean a large reduction of man's range or his numbers or both—and this is a catastrophe men will not accept voluntarily.)

The solution to the problems introduced by a technological advance is, *and always has been,* another step forward. The life of the farmer (peasant, peon, serf, slave) has never been ameliorated by the occasional wild rebellions in which they have engaged. What *has* improved their lot has been the discovery of new land, as in North America (a discovery made possible by the technics of the mariner's compass and the oceangoing ship) and, ultimately, by the nineteenth-century advances in the technology of agriculture.

Once farming was mechanized and chemical fertilizers were developed, fewer men with less work could produce far more food, so that it became less necessary to treat farmers with subhuman cruelty. And at the same time, the percentage of artisans, artists, intellectuals, entertainers, executives, clerks, and other nonfarmers could multiply still further.

As for the plagues of vermin and disease that the crowding of plants and men made possible, it is clear that these, too, were met with considerable success by the advance of technology.

It may seem that nontechnological factors are being given insufficient weight. What about that particularly nontechnological, yet particularly important, combination of ideas called religion?

Without trying to deny the importance of religion, it can be argued that it exerts its influence effectively only by means made possible by technological advance.

The earliest notions of religion antedate civilization. Neanderthal men buried their dead with elaborate ceremony and seemed to have notions of an afterlife. Nevertheless, how can a religion spread outward from its point of origin and truly influence large numbers unless it takes advantage of technology at least to the point of being written?

Is it entirely an accident that religious ideas originating with a few thousand tribesmen invading Canaan about 1250 B.C. should now be the basis of the beliefs of 1½ billion Jews, Christians, and Moslems? Or was it that, through the accidents of history, the Hebrews developed a group of superlative literary men who pro-

duced the books of the Bible, and that it was this written product, jealously preserved over the centuries, that kept the religion
alive and influential?

And could early Christianity have spread as it did but for the
historical accident of the existence of the Roman Empire? The
same ships and roads that carried the legions so well also carried
the Christian missionaries.

Consider the effects of a crucial advance in technology that is
closer to the present—that of printing, invented by Johann Gutenberg in about 1450.

Printing was the product of a vast number of developing technics—true paper, proper inks, movable type made of a metal that
expands on freezing, an effective printing press, and so on. It is
not surprising, therefore, that although the concept of printing is
extremely simple (much more simple, in my opinion, than growing plants or starting a fire) it took so long to develop.

Once invented, printing spread throughout Europe with, till
then, unprecedented speed, and had enormous consequences. For
one thing, it increased the number of copies of each book and
made it simple to publish even worthless ones. With the increase
in the numbers of books, it became less likely that some products
of the human mind might be lost. Hundreds of plays and philo-
sophic treatises produced by the great thinkers among the ancient
Greeks have been utterly and forever lost—but not one important
(or unimportant, I think) book has completely disappeared since
the invention of printing. This, in turn, meant that since 1450,
a "Dark Age" has become increasingly unlikely except through
the agency of truly titanic physical destruction, as in a thermonuclear war.

Then, too, the presence of many books meant the growth of
literacy. In an age when there was scarcely anything to read,
there was scarcely any need to know how to read. With the spread
of the printed word, the pressure to know how to read intensified.
Literacy spread, education expanded. For the first time, it became
possible for the general population to participate in the decisions
reached by the rulers of a large political unit.

It was only after the invention of printing, which made cheap
books and newspapers available to the general public, that de-

mocracy became possible over any territory larger than a city-state. Not all the efforts of Washington and Jefferson could have made the United States a democracy without the advance in technology represented by Gutenberg's invention.

And here we can clearly see the influence of technological advance on religion. In 1517, Martin Luther nailed his ninety-five theses to the door of the church at Wittenberg, and that began the history of the Protestant Reformation. But there had been those who quarreled with religious orthodoxy in every century of the Middle Ages, and none before had been successful. Why did Luther's movement remain unsuppressed?

Whatever political and economic reasons might be brought out in explanation, we can also see that when Luther began his work, printing was well established; and Luther made use of it. He poured out a never-ending stream of vigorously written pamphlets that filled Germany from one end to the other, and in that way his ideas spread faster than they could be slapped down. For the first time, heresy was not an argument among monks—for an increasingly literate population was drawn into the balance, and Protestantism survived.

Then, too, it was the printing press that made modern science possible.

The ancient Greeks had produced scientific speculations, some of which were on an exceedingly high level. These were embodied, however, in manuscripts that could be copied only with great difficulty and existed in only small numbers. No more than the few educated could know of them, and in a minor catastrophe, such as the sack of a city, the entire scientific product of a culture might be destroyed—and, in fact, this was what happened when the Crusaders sacked Constantinople in 1204.

Once printing was developed, the scientific thinking of men could spread outward wholesale and could neither be suppressed nor destroyed. When Copernicus' manuscript placing the Sun and not the Earth at the center of the planetary system was published in printed form, it was scorned, denounced, and thundered against; but it could not be made to vanish and, eventually, it won.

In fact, as scientific speculations became easily available in quantity, there arose for the first time what we might call a "community of science." Each scientist, wherever he might be in Europe, could climb upon the shoulders of his predecessors. It is no accident that what we call "The Scientific Revolution" took place on the continent that saw the printing press invented and in the century after the invention.

But if science is the daughter of technology, it quickly repaid its debt. Through science, men learned with increasing rapidity and increasing depth the nature of the laws of the universe, and by the proper understanding of those laws, it became easier to apply those laws to practical needs. So the advance of technology moved forward with enormously increasing speed.

The interworking of science and technology reached a kind of dramatic climax in the eighteenth century. Iron ore mixed with charcoal from wood produced iron when heated. Increasing need for iron and the declining supply of wood in Great Britain made it necessary to substitute coal from the earth. For coal to be mined, the mines had to be pumped dry of water.

Well, the importance of heat in forming metals led scientists to study the phenomenon from the theoretical standpoint. The Scottish chemist Joseph Black, studying heat in 1764, developed an understanding of the interconnection of heat with the boiling of water and the condensing of steam.

Black's friend, the Scottish engineer James Watt, used Black's theories to work out a design, in 1769, for the first practical steam engine that could use the power of expanding and condensing steam to activate a pump that would remove water from coal mines, and pump pistons and turn wheels in a variety of ways.

Watt's steam engine was the first "prime mover" based on heat. It was the first device that took heat energy from the inanimate world (burning coal, for instance) and put it to work doing things that until then it was customary for animal muscle to do. The steam engine powered machinery in factories, drove locomotives over steel rails, and guided ships over water.

With the steam engine there began the "Industrial Revolution," which completely altered the world. From 1800 to the present time, the population of the world has increased fourfold, and unexampled wealth and comfort have poured down on a larger and larger percentage of that increased population.

It brought its problems, as agriculture did, and as fire did, and there may be a longing for a pre-industrial "simple life"; but we cannot retreat to that any more than a farming community could retreat to food gathering.

We now have the problems of overpopulation, of pollution, of diminishing resources, of the risk of totally destructive war-weapons. But what is the solution? A retreat from technology? Impossible.

As always in the past, this can only be done by decreasing population and restricting range. Mankind has never agreed to do this voluntarily and never will. This can only be done by giving up the benefits of our technology, and mankind will not do so.

The solution, then? Why, what it has always been—the still further advance of technology.

Hand in hand, science and technology can find new and unlimited sources of energy, clean and safe, and with such energy, we can clean the world, recycle its resources, and reduce its inequities.

Hand in hand, science and technology can study man's psychology, behavior, and reproductive physiology, and find some humane way of reducing the birthrate and keeping the population from dangerous increase.

But can we be sure that science and technology will find these answers? Can we be sure that solutions to our problems exist? No, we can't be, but we can be sure that nothing but science and technology can find them if they do exist.

Can science and technology advance quickly enough to find the solutions in time? We can't be sure of that, but we can be sure of this: If mankind, generally, allows itself to be disillusioned and to turn against science and technology, it will succeed in making it certain that the advance will *not* be quick enough and that civilization *will* be destroyed.

To put it as briefly as possible: Science and technology are the answers to our problems. If they are not, nothing else is.

2 · Technology and the Rise of the United States

In 1783, when the United States, having survived the Revolutionary War and having won its independence from Great Britain, took its place among the powers of the Earth, its future was assessed by a shrewd and worldy wise monarch. Frederick the Great of Prussia, who had astonished Europe with his military victories a quarter century before, dismissed the new nation as a mere temporary freak. It could not exist for long, he said, because it was too large; it would fall apart.

In a way, Frederick was right! The new nation was 880,000 square miles in area—four times as large as France—and had a population of only a little over 3,000,000 squeezed into the Atlantic coastal area. Except for certain patches of that coastal area, it was a trackless wilderness. Even where roads existed, the best stagecoaches, traveling eighteen hours a day, took three days to go from New York to Philadelphia. To reach one corner of the nation from another took weeks. Although George Washington was supposed to be inaugurated as first President on March 4, 1790, the difficulty of travel from Virginia to New York delayed the swearing-in till April 30.

Under those circumstances anyone would guess that the nation was too large to hold together. Different portions, each of manageable size, would go their own ways—provided, that is, that those circumstances remained as they were. But they didn't!

It was the good fortune of the United States to have begun its life as an independent nation just at a time when technological advance was going into high gear, and to have drawn its cultural inspiration from the nation (Great Britain) in which that advance first went into high gear.

From the very start, the new nation had appreciated the importance of technological advance. Indeed, one important advance of this sort had helped make independence possible.

Although it was fashionable in cultivated European circles to laugh at Americans as backwoods barbarians and uncouth hillbillies, one American had gained the respect of the whole world. That was Benjamin Franklin. Franklin studied the properties of electricity, showed that lightning was of the same nature as the electricity produced in the laboratory, and, in 1752, applied his knowledge of earthly electricity to the lightning and developed the lightning rod.

The lightning rod spread quickly over the United States and Europe and proved, at trifling investment, to protect against the heavenly bolt of electricity. It was the first time a natural disaster had come under human control through scientific study and technological application, and Franklin was lionized. When, during the Revolutionary War, he came to France to negotiate for the French aid without which the embattled colonies could scarcely have won the war, it was his reputation as tamer of the lightning that, in part, helped to win over French public opinion.

The new nation lacked, at first, the industrial know-how that characterized Great Britain, which was the world's technological leader at this time (partly, this was because Great Britain's colonial policies had more or less deliberately held back American industrial development in the interest of the British economy). Nevertheless, the United States was quick to borrow.

In 1789, an Englishman named Samuel Slater arrived in the United States. He had worked in those British factories that were beginning to use the steam engine to power devices that spun threads and wove cloth, replacing the slower hand labor. It was this that marked the beginning of the "Industrial Revolution." Slater had the designs of such machinery in his head and, in 1790, established a powered factory in Pawtucket, Rhode Island. In this way, the Industrial Revolution came to the United States.

Oliver Evans of Delaware was building high-pressure steam engines before 1802, and with these, factory after factory could be powered. Francis Cabot Lowell of Massachusetts built elaborate spinning and weaving mills. The process of American industrialization had begun, and it was to continue unabated ever since.

Many Americans, such as Thomas Jefferson, viewed industrialization with alarm, feeling that virtue and happiness were to be found only in a rural society of small, independent farmers. Without arguing the morality of the situation, we might say that in all history increases in standards of living have always accompanied industrialization and no society has ever voluntarily accepted a decline in those standards and abandoned industrialization to return to a farm economy.

Moreover, the Jeffersonian dream of a society of hard-working, virtuous farmers assumed a continuation of conditions as they were in early America—a great deal of cheap land and a small population. As population increased, a purely farming society was bound to become impoverished, and there is little virtue to be found in starvation. The nations that today are predominantly agricultural are, in general, poor and miserable.

In an industrializing society, farming itself becomes subject to technological advance, and in this, the United States actually led the way from the very start.

Cotton was one of the important crops of the early South. It was in great demand in Great Britain which, in steam-powered factories, was producing cotton cloth in greater quantities and more cheaply than ever before.

The southern states could easily sell far more cotton than they produced, but the bottleneck came in the plucking of cotton fibers from the seed—a dreadfully tedious and slow job.

A young Connecticut gadgeteer, Eli Whitney, visiting Georgia in 1793, heard planters complaining about the difficulty of plucking cotton fibers. He devised a simple spiked cylinder which, when it rotated, entangled the fibers and pulled them off the seeds mechanically.

The "cotton gin" (short for "engine") increased the rate at which cotton could be picked off the seeds at least fifty fold. At once it became possible to grow far more cotton, since more people could be placed in the field, and even large cotton crops could be quickly handled by building more cotton gins.

(Technology has unforeseen side effects. Whitney could not have known that the expanding cotton fields would increase the South's use of black slaves. Slavery, which seemed on the point of dying out by then, became so important a part of the southern economy that eventually the southern states went to war rather than give it up. Yet it was not the cotton gin itself that fastened slavery on the South, but the mistaken human decision that only slaves could cultivate the cotton fields. Free labor could have done even better, and did do so after slavery was abolished. And more machinery would do still better.)

Whitney made no money out of the cotton gin. It was so simple to build and use that anyone could infringe on the patent, and everyone did. Whitney returned to Connecticut and there, in 1798, turned to the manufacture of firearms. He did this with precision and was the first to manufacture parts so nearly identical as to make any part fit any gun. This was even more important, in the long run, than the cotton gin.

But American inventiveness had to be applied to the problem of making the nation less unwieldy in terms of transportation and communication. The problem was intensified by the fact that in 1803, the United States had bought the vast territory of Louisiana, another virtually trackless wilderness. This doubled the American area to 1,700,000 square miles—as large as all of Europe outside Russia.

New and better roads were built. The "turnpike" (or toll road), a paved highway maintained by the government, had been introduced by Great Britain and was quickly adopted by the United States. The first American turnpike was a sixty-two-mile stretch from Philadelphia to Lancaster, Pennsylvania, opened to traffic in 1791. By 1810, there were three hundred turnpike corporations in the Northeast.

Canals were also built, and these interlaced the settled areas. The longest and most ambitious canal was the Erie Canal, which carried ships from the Great Lakes to the Hudson River. It was opened in 1825 and brought the American interior to the Atlantic Ocean by means of a waterway. New York City, which was the ocean port to which the products of the interior converged by way of the Great Lakes, the Erie Canal, and the Hudson River, quickly became the largest city in the United States and, eventually, the richest and most remarkable metropolitan area in the world.

The process of road and canal building would have progressed more quickly and benefited America more greatly if it had been placed under federal control. Through the first half of the nineteenth century, however, Jeffersonian ideas won out and it was the individual states (who could not always afford it, and who did not always co-operate) that were in charge.

But transportation had not merely to be facilitated; it had to be improved. Might not the power of the steam engine be applied to the turning of a wheel that would drive a ship against the wind and current? As early as 1787, John Fitch of Connecticut had built, and was running, a steamship on the Delaware River. Bad breaks drove him into bankruptcy, but in 1807, Robert Fulton, born in Pennsylvania, had better luck on the Hudson River. The steamship, independent of wind and current, knit the nation together by sea and river.

A similar device intended to run a vehicle on land over iron rails (a "railroad") was worked on by many people, both in Great Britain and in the United States. The first American steam locomotive capable of drawing after it a train of cars containing merchandise or people, was built in 1830 by Peter Cooper of New York.

A veritable explosion of railroad building was set off which, in the course of a single generation, knit vast sections of the nation together by means of thousands of miles of shining rails along which the trains steamed. For the first time in history, travel by land was as rapid and convenient as travel by sea. With that,

Frederick the Great's cynical prediction no longer had any chance of coming true. At least if the United States fell apart, it would not be merely through unwieldiness. Technology had succeeded in making size alone no danger.

Technology dealt not only with the transportation of men and goods but also with the more subtle transportation of ideas—in a brand-new way.

Through the nineteenth century, the United States, for all its skill at technology, was poor in scientists. But it had some, and one of the most important was Joseph Henry of New York who, in the 1830s, made important fundamental advances in the knowledge of electricity.

Making use of Henry's discoveries, Thomas Davenport built the first electrical motor in 1834. Then, too, Samuel Finley Breese Morse of Massachusetts, an artist by profession and a promoter by inclination, devised (with Henry's help) an electric telegraph whereby signals could be sent over long distances. He managed to persuade Congress to provide the money and built a telegraph line from Washington to Baltimore.

In 1844, the first message went winging along wires at the speed of light. That message, making use of a biblical phrase, was "What hath God wrought?"

It meant that one era had come to an end, and another—that of instantaneous communication—had arrived. The Battle of New Orleans had been fought in 1815 and thousands had died, even though peace had come three weeks before—for the news had not crossed the Atlantic Ocean yet. But in 1858, telegraphic communication by means of a cable laid across the Atlantic by the American financier Cyrus West Field of Massachusetts, made it possible for Queen Victoria to talk to President Buchanan as though the three thousand miles between them had disappeared.

When the American Civil War began in 1861, Frederick the Great's prediction met its ultimate test and proved false. It was technology that saved the Union. The North, because it had industrialized, was far richer and economically stronger than the still rural South. The telegraph made it possible to control the armies in the field at any distance. Most of all, the railroad network in the North (far more advanced than that in the South) made the northern armies more mobile and easier to supply than those of the South.

Without its inferiority in industrial and technological progress, the South, fighting with great bravery under the best generals America has ever produced, could not have been beaten, and the United States would indeed have fallen apart.

By the time of the Civil War, the United States (quite un-noticed by a Europe that still tended to view it as a backward and barbarous country) had taken the technological lead in the world in almost every direction. New inventions, each with in-calculable effects on American society, poured out of ingenious American minds.

In 1831, Cyrus Hall McCormick of Virginia invented a mechani-cal reaper, which introduced the great movement toward mech-anized agriculture; one in which machines (eventually powered by gasoline engines) took the place of human laborers. This meant that fewer and fewer farmers, controlling larger and larger supplies of energy, could produce more and more food.

In 1800, nearly everyone in the United States was a farmer, and they produced very little more food than that required to feed themselves. By 1973, fewer than one in twenty Americans lived on farms, and they produced enough food to feed the other nine-teen lavishly, with still more to send abroad.

It was the mechanization of agriculture that made it possible for so many Americans to engage in industrial, commercial, and service occupations and to make the United States the most ad-vanced technological nation in the world.

Other American inventions peppered the pre-Civil War years. Samuel Colt of Connecticut invented the revolver, or "six-shooter," in 1835, which was, as we all know, an important adjunct in the winning of the West.

The Swedish inventor John Ericsson had, in 1836, invented the screw propeller for use in steam ships, replacing the large and vulnerable water wheel. It was the screw propeller that made steam warships practical. Foreign inventors, however, were coming to the United States in greater and greater numbers, for in the new nation, the social atmosphere was particularly favorable to technological innovation.

Ericsson arrived in the United States in 1840 and worked out a number of devices that served to increase the efficiency of warships. His most important invention was that of the iron-clad warship, *Monitor* which steamed south in 1862 just in time to take on the southern iron-clad, *Merrimack*. By neutralizing the latter, the *Monitor* prevented the South from breaking the Union blockade and, perhaps, the Union with it.

The European powers affected to despise the American Civil War as the mere barbarous clashes of large masses of men. (They failed to understand that the Civil War was the first modern technological war and that, had they studied it more closely, they might have avoided many of the mistakes of World War I.) The battle of the *Monitor* and the *Merrimack*, however, they could not ignore. Great Britain realized, suddenly and uncomfortably, that every wooden warship was now potentially useless and that entirely new iron-clad fleets would have to be built.

For the first time, the United States clearly and admittedly showed the way in the application of technology to warfare.

But there were peacetime advances, too. In 1839, Charles Goodyear of Connecticut discovered (accidentally) the process of the vulcanization of rubber, which, for the first time, made rubber a practical industrial material. And in 1859, Edwin L. Drake of New York was the first to drill for oil, near Titusville, Pennsylvania. The time was to come when rubber and oil were to combine to make the automobile possible.

In 1846, Elias Howe of Massachusetts built the first practical sewing machine and this, for the first time, brought technology into the life of the housewife and began a process that liberated women from age-old oppression more effectively than all the humanitarian speeches ever made.

In 1852, the first practical elevator was devised by Elisha Graves Otis. It was this that was eventually to make the skyscraper possible.

Inventions as humble as the safety pin, invented by Walter Hunt in 1849; as deadly as the repeating rifle, invented by Oliver F. Winchester in 1860, and the machine gun, invented by Richard

J. Gatling in 1862; as impressive as the rotary printing press, invented by Richard M. Hoe in 1846, which made the modern large-circulation newspaper possible; poured forth.

A number of factors combined to make the United States of the latter half of the nineteenth century the clear technological leader of the world. The very fact that it was immensely large and underpopulated meant there was no fear of machinery replacing manpower—there wasn't the manpower to replace.

The fact that it was a nation of immigrants, with people pouring in from all over Europe, meant that there was a mixture of cultures that prevented the establishment of one particular way of life that was considered too hallowed to change. That, combined with the new ground being constantly broken farther and farther west, produced a kind of atmosphere of ever-new that made the very concept of change seem an absolute good.

Furthermore, the United States, throughout the nineteenth century, was isolated from the cockpit of Europe and was secure behind a more or less friendly-to-indifferent British fleet. This meant that the United States could concentrate on the arts of peace. No foreign army landed anywhere on American shores after 1814, and America could sharpen its military technology in brief, nondangerous wars with Mexico and the Indians. The one great war it fought, the Civil War, was largely fought on southern rural territory, and the northern industrialized area remained untouched and unharmed.

The result of all this was that in the decades after the Civil War, the whole world had to come to the reluctant realization that technological leadership had passed to the crude and vigorous Americans. When Jules Verne wrote his novel about the first flight to the Moon, in 1865, it was not the great nations of Europe whom he envisaged as organizing the conquest of space. No, it was Americans whom he saw as manning the first spaceship to the Moon—and here he was, as in so many other cases, right.

But if there was one man who burst forth as the genius of invention, as the embodiment of technological advance, as the very symbol of American leadership in the new movement, it was

Thomas Alva Edison, born in Ohio in 1847. In 1869, he devised
a stock ticker better than any that existed. Fearful of asking five
thousand dollars for it, he asked for an offer and was offered
forty thousand dollars. He was in business.

At the age of twenty-three, he founded the first firm of con-
sulting engineers and for the next six years turned out inventions
such as waxpaper and the mimeograph, working twenty hours a
day and sleeping in catnaps.

In 1876, Edison set up a laboratory in Menlo Park, New Jersey,
the first industrial research laboratory. Eventually, he had as many
as eighty competent scientists working for him, and that was the
beginning of the modern notion of the "research team."

Before he died, he patented nearly thirteen hundred inventions,
a record no other inventor has ever approached, and in one four-
year stretch he obtained three hundred patents, or one every five
days. He was called the "Wizard of Menlo Park," and in every
corner of the world the American inventor was held in awe.

In Menlo Park in 1877, Edison improved the telephone, invented the year before by the Scottish-American inventor Alexander Graham Bell. Edison also produced the phonograph and,
in 1879, reached the peak of his career with the invention of the
electric light. So enormous was his reputation that when he merely
announced that he would *try* to invent an electric light, illu-
minating-gas-company stocks tumbled at once in the stock ex-
changes of New York and London.

Later still, Edison invented the motion picture and discovered
the "Edison effect," which he did not follow up, but which served
as the basis for later development of electronic devices such as
radio and television.

So as 1900 approached, the United States had become the land
of technology par excellence. Its railroad network was the largest
and best in the world and spanned the continent; its standard of
living was rising steadily; its cities were moving upward as well
as outward; its business was expanding with limitless confidence.

It lacked a sizable army and a military tradition, and the "great
powers" of Europe still counted greatness in terms of armies,
navies, and far-flung colonies. But then, when the United States,

in 1898, defeated Spain (not one of the great powers, to be sure) with ridiculous ease and took over island possessions in the Atlantic and Pacific, the United States was granted the status of a "great power."

In the twentieth century, the airplane, the nuclear bomb, television, and rocketry were all fields in which the United States took the leadership, and the gap between American technology and those of all other nations steadily widened.

American military force was clearly equal to that of the battered European powers in World War I, which she entered in 1917, unprepared; and clearly superior to them in World War II, which she entered in 1941, again unprepared. After World War II, she was the stronger of the two "superpowers" and without doubt the strongest and richest nation in the world.

The American rise to power clearly paralleled her rise in technology. In telephones, in automobiles, in television sets, in every aspect of modern living, she leads the world. In agricultural production, she leads the world, too. While it might be said that material goods are not necessarily to be equated with happiness, no one has ever said that the lack of food, clothing, and shelter leads to happiness either.

If there is a certain garishness to the American way of life, it may also be said that American security, developed first through isolation, and then through the realization of power, has enabled the United States to retain a high degree of political and civil freedom.

Freedom is a luxury, and only the United States, of the world's great powers, can afford to follow policies, both foreign and domestic, that require strength and determination, while still allowing portions of its population the freedom to oppose those policies openly and loudly. Other nations lack the solid technological background to risk the weakening of purpose that such opposition can bring and must paper over their insecurity by forcing a brittle conformity upon their people.

But there is a catch. Almost every step forward that the United States has made in its developing technology has been at the cost of increasing the rate of its energy consumption. Until World

War II, it has been able to do this at the expense of its own (fortunately) immense supplies of wood and waterpower, then coal, and then oil. At the present moment, the United States is making use of some 40 percent of all the energy being expended on Earth, but it is beginning to grow dependent on foreign sources for oil—today's most convenient energy source.

If the American source of strength is not to be delivered into other, and possibly unfriendly, hands; if the United States is not to present the world with an exposed throat that can be cut at the whim of another—then new energy sources must be sought at home. Coal, which is still plentiful, must be used more safely and efficiently; nuclear energy must be safely used, and nuclear fusion must be made practical; geothermal energy and direct use of solar energy must be developed.

In whichever direction we look, however, we see, plainly and unmistakably, that American strength, American comfort, and American liberties can be maintained and advanced only by a studied and wise continuing advance in technology.

Technology, which turned the United States from a wilderness few suspected could be held together, into the richest, strongest, and most free nation that ever existed on Earth, can continue to keep the United States in that happy position. To turn away from technology is to lose what, for two centuries, has been the last best hope of mankind.

Chapter 2 • AFTERWORD

The first two essays in this book brought me an unusual pleasure, and I must tell you about it.

I was asked to do the essays by a Chicago engineering firm which, alarmed by the spreading distrust of technology among the population, wished to undertake a series of pamphlets giving the case for technology. Since I, too, am alarmed by the spreading distrust of technology among the population, I quickly agreed to do the first two pamphlets, which were designed to set the stage for the remaining ones.

I was a little taken aback when I was told that my name would not appear on the pamphlets I wrote, but after some thought I decided the importance of the task warranted the violence done my egocentricity, and I agreed to accept anonymity.

It did not occur to me at the time that I was setting up a situation at last that might help bring an answer to a question that had often plagued me.

As you know, I write a great deal, and I very frequently get reviews (or other comments) that say nice things about me and that use such adjectives as "lucid," "clear," "enthusiastic," "sprightly," and so on, and so on.

Naturally, I just love words of that sort, but after a while I grew uneasy. Perhaps it was all a conditioned response. By now, perhaps, the button labeled "Asimov" automatically brought up the adjectives, and maybe they had no real meaning.

Sometimes I would brood and wonder if I ought not arrange

to have something published without my name on it, just to see what would happen. Frankly, I never quite dared. I was afraid I might get something like this, "Though the writer, whoever he may be, is quite obviously imitating Asimov's style, he fails miserably and is merely confused, muddy, boring, and dull." I didn't think I could bear to have that happen.

But then these two articles came out—anonymously—and the engineering firm sent copies to all kinds of important people, looking for quotations they might use in promoting their project. They received numerous responses, and they sent copies of those responses to me.

And there they were! All the adjectives I was used to seeing! And no one knew it was I who wrote the essays!

I tell you I didn't stop grinning for forty-eight hours.

Anyway, here they are now under my own name.

Chapter 3 • FOREWORD

As you may or may not know, magazine editors sometimes change jobs. And it frequently happens that when one does, he carries me from post to post like a virus disease, until I now infest virtually the entire magazine field.

For instance, an editor who had worked with *Science Digest*, and for whom I had done some work, took over the post of editor of *Modern Maturity*, an excellent magazine serving the senior citizen.

The editor thought it might be interesting to have someone write short articles detailing the changes that the twentieth century had brought in this field or that—since the period from 1900 on represented the life span of the average reader of the magazine.

In this connection, he thought of his old writer-acquaintance, Asimov, and called me up to ask if I could do such a series.

"Of course," I said, propping the telephone against my shoulder and placing a sheet of paper into the typewriter. "How many?"

I ended up doing nine of the articles, which in this book make up Chapters 3 to 11, inclusive. I would have done more, but he said nine was enough. I think he would have stopped me sooner, but he didn't get the words out fast enough.

3 · The Transportation Revolution

The twentieth century has seen two enormous revolutions in transportation, moving along on parallel tracks: one on the ground, one in the air.

When 1900 opened, the most advanced form of long-distance land travel was the steam locomotive railway train; of long-distance sea travel, the steamship—each a prime example of the nineteenth-century revolution in transportation.

For day-to-day short-distance travel on land, however, there remained the horse, with a human being either astride the saddle on its back, or in the buggy or carriage behind. The horse, whose use in transportation had originated in central Asia in 2000 B.C., was still man's prime means of covering ground nearly four thousand years later.

There were "horseless carriages," to be sure, or "automobiles" ("self-movers"), as they came to be called, which were driven by the inanimate power of a steam engine or a gasoline engine. In 1900, however, these were still little more than toys for rich men in search of novelty. There were only about eight thousand of them in the nation, altogether.

A Detroit mechanic named Henry Ford had built his first car in 1896, but it was not till 1903 that the Ford Motor Company began to offer automobiles to the public. In 1908, Ford began to produce the Model T, making use of assembly-line techniques eventually, and for the first time the automobile appeared in large enough numbers and at low enough cost to be within the reach of the average man. The Model T cost $850 when it first appeared, and the price had come down to $265 by 1923.

In the opening decade of the twentieth century, the automobile engine had to be started by hand cranking, which was hard work, and dangerous, too. If the engine caught, and the crank got away from you, it could, and sometimes did, whirl around and break your arm. That, more than anything else, limited the use of the Model T.

In 1912, Charles F. Kettering invented the first practical battery-powered self-starter and placed it in the Cadillacs he was building. Once the self-starter became common, it meant that women and youngsters could drive cars easily. There was no longer a premium on muscle and nerve.

By 1920, the United States was clearly on its way to becoming a nation on wheels. There were 6,000,000 automobiles on the roads.

The revolution that had thus begun continued to spread and intensify for half a century. The horse has all but disappeared as, with each year, city streets and country roads become more alive with automobiles, trucks, and buses. The number of automobiles now registered in the United States is in excess of 125,000,000; and in the world over, more than 250,000,000.

America's love affair with the automobile has brought a host of subsidiary revolutions in its wake. The need for gasoline has enormously altered man's scheme of energy production and has done more than anything else to spark the changeover from coal to oil. Rubber has become a vital need because of its use in tires. New, broad highways form a network over the nation and are lined with motels and gasoline stations, while the cities themselves are filled with garages and parking lots.

All this has made Americans a nation of travelers. It has made vacations more elaborate and has helped break up the close-knit family group. The greater freedom for women and the more tolerant attitude toward sex that began in the 1920s, and has been growing ever since, was due, at least in part, to the ability of women to travel by themselves and to the ease with which young people could arrange, via automobile, to be in each other's company away from home.

New problems have arisen. Traffic jams choke city centers to the point where walking becomes the fastest mode of travel over moderate distances. Smog, to which gasoline-engine exhaust is a major contributor, hangs like poison gas over America's major cities. What's more, fifty thousand Americans are killed each year in traffic accidents.

But let us leave the ground.

When 1900 opened, no human being had ever moved through the air for extended periods independently of the wind. For over a century, men had floated through the air, suspended by hydrogen-filled balloons, going where the wind blew. For some decades, men had ridden the air currents in light, wooden gliders, taking skillful advantage of moving air, but still remaining slave to it.

Before the year was half over, however, the German inventor Ferdinand von Zeppelin had placed balloons within a cigar-shaped aluminum shell. (Methods for producing that light metal cheaply and in quantity had only been worked out a dozen years before.) Beneath the shell was a gondola, carrying engines that turned propellers. On July 2, 1900, the first "dirigible balloon" (one that could be "directed" even against the wind) flew, and powered flight was a reality.

For over thirty years these large dirigibles were built. They crossed the ocean with loads of passengers and they circled the planet. They were too fragile, however; too easily set afire if hydrogen was used; too easily broken up by storms even if non-flammable helium was used.

The future of powered flight lay with the glider, as it turned out.

On December 17, 1903, the American gliding enthusiasts Wilbur and Orville Wright placed an engine on a glider and successfully flew 120 feet, remaining in the air for 12 seconds. That was the first powered flight of a heavier-than-air machine.

Other "airplanes" were built. In 1905, one of the Wright

brothers stayed in the air half an hour and flew 24 miles. In 1909, the French aeronaut Louis Blériot flew from France to England, across the English Channel, in a home-built plane—the first international flight.

As late as 1914, though, airplanes were still little more than engines mounted on gliders, with daredevil stuntmen riding them. But then came World War I, and it was not long before all sides saw the advantage of aerial reconnaissance and aerial bombing. The airplane grew to maturity in that war. It did relatively little damage in this first air war, however, and added the one note of glamor to what was otherwise a war fought with less imagination and poorer generalship than any in history.

On May 15, 1918, before the war had ended, the United States began its first attempts to establish airmail service between Washington and New York. By 1921, transcontinental mail was being carried by air.

More and more glamorous feats were performed by aeronauts. In 1919, two British pilots, John Alcock and A. W. Brown, made the first nonstop flight across the Atlantic, from Newfoundland to Ireland, in 16 hours. In 1923, two American pilots, O. G. Kelly and J. A. Macready, flew nonstop from New York to San Diego in 27 hours. In 1926, the Americans Richard E. Byrd and Floyd Bennett, flew over the North Pole.

It was, however, the solo flight of Charles A. Lindbergh, from New York to Paris, in 33½ hours, on May 20–21, 1927, that really caught the imagination of the world and convinced the public that air flight might be a practical method of public, large-scale transportation. By the mid-1930s, the Douglas DC-3 was routinely carrying up to 21 passengers at speeds of up to 180 miles an hour.

Larger and larger planes were built and speeds of up to 400 miles an hour were attained, but the weak point of the plane was the propeller. It could only turn so fast and it could pull the plane only forward. Again it took war, World War II, to accelerate new developments.

In one direction, the airplane was made more maneuverable by placing a large propeller on top. The first commercial "helicopter" was designed by Igor Sikorsky in 1940. In addition to flying forward as an ordinary plane could, it could ascend and descend vertically, it could hover, and it could move backward. It could not move quickly, however.

For speed, propellers were abandoned altogether, and the action-reaction principle was used. Fuel was burned, and the exhaust gases, driven fiercely backward, forced the plane itself forward. All the major nations were experimenting with such jet planes during and just before World War II, but the first really practical one was designed by a British Air Force officer, Frank Whittle, in 1941.

After World War II, the jet plane became the primary mode of carrying passengers long distances at unprecedented speeds. Planes now carry people by the hundreds rather than by the dozens. On October 14, 1947, a jet plane first flew faster than the speed of sound, 760 miles per hour. Within a few years, such "supersonic speeds" became routine.

As the automobile after World War I made short-distance travel common, so the jet plane after World War II made long-distance travel common. The whole world has become the playground of the "jet set" and has become the single arena of the businessman.

Between the two, the automobile and the jet plane, those great transportation devices of the nineteenth century, the railroad and the ocean liner, have withered and become obsolete.

Even the jet plane is not the ultimate. It burns fuel in oxygen, which it sucks in as part of the surrounding air. It can only fly where the atmosphere is thick enough to support combustion.

In the 1920s, however, the American physicist Robert H. Goddard was experimenting with rockets that carried not only fuel but also liquid oxygen in which that fuel might burn. Such rockets are independent of air and can travel even in the vacuum of outer space.

In 1926, Goddard sent up his first rocket near Worcester, Massachusetts. In the early 1930s, working in New Mexico, he succeeded in sending rockets 1½ miles up into the air, and watched them attain speeds of up to 550 miles per hour.

Goddard received no government support, but similar rocket experimenters in Germany did. Under the direction of Wernher von Braun, V-2 rockets were developed, and were used by the Germans to pound London in 1944.

After World War II, the United States and the Soviet Union flung themselves furiously into rocket research. In 1957, the Soviet Union sent the first man-made object into orbit about the Earth at a speed of 5 miles per second. The United States soon followed with a similar feat.

On April 12, 1961, Yuri Gagarin of the Soviet Union became the first man to circle the Earth beyond the atmosphere, and on July 20, 1969, Neil Armstrong of the United States became the first man to set foot on another world, when he stepped out onto the surface of the Moon.

The century began, in short, with no powered air-flight of any sort having ever been made by man, and before seventy years had passed, man had lifted himself to the Moon.

It all happened in the space of a single lifetime. Individuals who might have seen the first dirigible move through the air in 1900, if they had been in the right place, lived to see, via television, men walking the Moon.

4 · The Atomic
Revolution

In 1900, scientists were confronted with a problem concerning the atom. All through the nineteenth century, atoms had been viewed as very tiny, featureless balls and were thought to be the smallest particles that could possibly exist. Each different element (gold, iron, aluminum, oxygen, and so on) was considered to be composed of a characteristic kind of atom, different from those of any other element. All the atoms of one particular element were held to be the same.

In the 1890s, however, a kind of particle called the "electron" had been discovered, and it was far smaller than even the hydrogen atom, which was the lightest of all atoms. At the same time, it was found that the two elements with the heaviest known atoms, uranium and thorium, were "radioactive." They gave off puzzling radiation that might be composed of tiny "subatomic particles," also far smaller than atoms.

As the twentieth century opened, scientists concentrated on the study of radioactivity. It turned out that when a radioactive atom gave off its radiation, it changed into another kind of atom, which then broke down to still another kind, and so on. In 1904, the American physicist Bertram B. Boltwood demonstrated that there was a whole chain of such breakdowns and that uranium atoms, breaking down through numerous steps, finally became lead atoms.

This happened very slowly. Of any supply of uranium, half turns to lead only after 4.5 billion years. Boltwood suggested, in 1907, that rocks might be analyzed for both uranium and lead. By judging how much of the former had turned into the latter, the age of those rocks could then be measured. By this method, geologists got an accurate idea of the age of the Earth's rocks for the first time, and eventually they were quite confident that the Earth was formed about 4.7 billion years ago.

The radiation that uranium gives off contains energy. Where does all the energy come from? Only a very little is given off over the space of an hour or even a year, but that little mounts up greatly over the billions of years. In 1905, the German physicist Albert Einstein worked out a "theory of relativity" based on the assumption that the speed of light was always the same.

By means of this theory, Einstein deduced that matter could be turned into energy and that a very little matter could be turned into a great deal of energy. The energy produced by radioactive uranium arose because a very small fraction of the uranium atoms themselves were turning into energy. Einstein's equation representing this was $e=mc^2$, where e stands for energy, m for matter, and c for the speed of light.

The New Zealand-born physicist Ernest Rutherford used the radiation of radioactive material to study the structure of atoms. He let some of the subatomic particles in the radiation strike a thin film of gold and was prepared to study the way they bounced off the atoms.

To his surprise, he found that most of them went straight through the gold as though there were nothing in the way. Every once in a while, though, some of the subatomic particles bounced off to one side. Very occasionally, some bounced directly backward.

Rutherford decided that there was hard, heavy material in the atoms off which the subatomic particles could bounce. This hard, heavy material, however, must take up a very small volume at the very center of the atom so that the subatomic particles didn't manage to hit them and bounce very often. On the outskirts of the atom, there was only light material that could not stop the subatomic particles.

The tiny volume of heavy material at the center was the "atomic nucleus." Around it, distributed through the rest of the atom, were light electrons. Each element had a different number of electrons, from only 1 in the hydrogen atom to 92 in the uranium atom.

Inside the atomic nucleus were a number of heavy subatomic particles. One type was called the "proton." There were as many protons inside the nucleus as there were electrons outside. The other type of heavy particle in the nucleus was not identified till 1932. It was then discovered by the English physicist James Chadwick, and it was named the "neutron."

As early as 1913, the English physicist Frederick Soddy had discovered that despite what had been thought in the nineteenth century, all the atoms of a particular atom were not alike. Sometimes an element consisted of two or more varieties of atoms, which were alike in all respects except that each variety was a little different in weight from the others. Each variety of atom that was present in an element was called an "isotope."

It quickly turned out that the difference in isotopes rested in the atomic nucleus, and the puzzle was solved after the discovery of the neutron. All the atoms of a particular element had the same number of protons in the nucleus, but the number of neutrons might be different. Thus, every uranium atom had 92 protons, and most of them had 146 neutrons also. The nuclei of such uranium atoms had 238 particles altogether, so these were called "uranium 238." One out of every 140 uranium atoms had 92 protons and only 143 neutrons. These were "uranium 235."

These isotopes gave scientists a way of studying the complicated chemical reactions in living tissue. Chemicals could be prepared with one of the rare isotopes of oxygen, nitrogen, or hydrogen, elements that occur in living tissue. Cells absorbed these chemicals and changed them. Scientists then determined which new chemicals had the rare isotopes and could, in this way, trace the chain of chemical changes that had taken place within the cell.

One way of studying the structure of atomic nuclei was to bombard them with subatomic particles (as Rutherford had done). When the particles strike the nuclei they might add on to them, or knock something out of them. Either way, that would change one isotope into another. In 1934, a French husband-and-wife team, Irène and Frédéric Joliot-Curie, discovered that isotopes could be formed in this way that weren't found in nature. Such isotopes had too many neutrons (or too few) for the atomic nuclei to be stable. They were therefore radioactive and gave off radiations of their own, changing into stable nuclei in this fashion.

These "radioisotopes" were even easier to use in studying the chemistry of living tissue than ordinary isotopes were. They could be used in particularly small amounts. Living cells cannot tell them from ordinary isotopes of the same element, and therefore incorporate them into various chemicals. The radioisotope can then be followed by the radiation it emits, and the finest details of the chemical changes in the cell can be worked out.

Once the neutron was discovered, it was quickly found that it was easier to hit atomic nuclei with this newly discovered particle than with any other. In 1934, the Italian physicist Enrico Fermi began to bombard the atoms of various elements with neutrons. Among others, he bombarded uranium atoms with neutrons. The results he got in this case puzzled him.

It puzzled others, too. In Germany, the matter was studied by Otto Hahn and by his Austrian-Jewish coworker, Lise Meitner. Meitner had to flee in 1938, after Nazi Germany annexed Austria and began to persecute Austrian Jews. Safe in Sweden, she thought about the work on uranium and decided that, when bombarded with neutrons, uranium nuclei did not simply add on a subatomic particle or lose one. Instead, it must split in two ("uranium fission"). This resulted in the development of much more energy than was true of ordinary radioactive changes.

This was announced in early 1939, and other physicists quickly checked and saw that this was the correct solution. A Hungarian physicist, Leo Szilard, who was a refugee from Nazi terror, noted that when the uranium atomic nucleus split in two, it released neutrons. These neutrons could split other uranium nuclei, releasing still more neutrons, which could split still other uranium nuclei, and so on. If enough uranium could be gathered in one place, and if certain conditions were met, such a "nuclear chain reaction" would release so much energy (in accordance with Einstein's $e=mc^2$ equation) as to produce a bomb far more violent than any explosive ever known to mankind.

Szilard urged other scientists to keep their work secret in order that the Nazis might not develop such a bomb first (for World War II was beginning). He and other physicists then persuaded Albert Einstein to write a letter to President Franklin Roosevelt urging him to start a project to develop such a bomb.

The "Manhattan Project" was begun, and on December 2, 1942, enough uranium was put together in a so-called atomic pile to start the first controlled nuclear chain reaction in history. By July 1945, scientists working under the direction of J. Robert Oppenheimer had put together uranium in such a way as to start an uncontrolled chain reaction at will. This was the first nuclear bomb, and it was exploded in Alamogordo, New Mexico.

By that time, the Nazis had been defeated, but Japan was still fighting the United States. A nuclear bomb was therefore exploded over the Japanese city of Hiroshima on August 6, 1945, and another over Nagasaki on August 8. This quickly put an end to the war.

Although enormous energies were produced by the fission of heavy atomic nuclei such as those of uranium, even more energy could be produced by forming the light nuclei of hydrogen to stick together to form heavier nuclei ("nuclear fusion").

After World War II, it was discovered how to use the energy of a nuclear bomb to make hydrogen nuclei to fuse and thus increase the force and danger of the explosion many, many times. Such a "hydrogen bomb" was first exploded by the United States on a Pacific island in 1952. Soon the Soviet Union, which had developed a fission bomb of its own in 1949, had fusion bombs as well. Great Britain, France, and China had also developed nuclear weapons since World War II.

So it has come about that whereas in 1900 virtually nothing was known about the interior of an atom, scientists had, within half a century, learned an enormous amount about the intimate details of its makeup. Whereas the century began with chemical explosions as the worst an army could do (the Spanish-American War, fought in 1898, made do with nothing more powerful than ordinary gunpowder), within half a century, the great powers of Earth had developed explosives of so fearsome a capacity as to endanger the survival of humanity.

Yet the research that led to the development of the monstrous nuclear bombs had other results, too. That very same research led to the development of methods for the formation of all kinds of radioisotopes in quantity.

These radioisotopes are now used in every kind of research—in engineering, in manufacturing, in agriculture. The telltale radiations of the radioisotopes can give hints and clues that can guide machines in factories, follow hidden flows of liquid, detect leaks and flaws in structure, control automatic operations—and, of course, continue to work out the secrets of life.

It is doubtful if the vast advances we have made in the past quarter century in understanding the complicated operations that control the way in which we inherit our characteristics from our parents, or the way in which plants use the energy of sunlight to produce food and oxygen, would have been possible without radioisotopes.

5 · The Energy
Revolution

In 1900, parts of the world had been industrialized for up to a century, and the energy demands, which increased constantly as industrialization proceeded, had been met by coal. Each decade more and more coal was mined—at great effort and at great danger.

There was a liquid fuel, too, called petroleum or oil, which could be drilled for and allowed to rise upward of its own accord. Men did not have to go underground to chip it out painfully. And oil could be sent, over land at least, through pipelines. Because it was liquid, its flow could be easily controlled and it could be used with precision.

As long ago as August 1859, the first oil well was drilled in Titusville, Pennsylvania, by Edwin Drake.

Before 1900, however, petroleum had been used chiefly for lighting. A high-boiling fraction, kerosene, fed the lamps of the nation. Once the electric light was invented in the 1870s, it might have seemed that even that use would die out.

But with the opening decades of the twentieth century, airplanes and automobiles were developed. Those were powered by internal-combustion engines within whose cylinders mixtures of flammable vapor and air were exploded over and over to keep the pistons pumping and the wheels turning. The cheapest and most useful fuel for the purpose was a low-boiling fraction of petroleum called gasoline. This could be transported and stored as a liquid, but vaporized easily within the engine.

Larger and larger quantities of gasoline were used as the world became motorized, and higher-boiling fractions of petroleum were used in the diesel engines developed for ships and trucks. By the time of World War I, petroleum had become a vital resource for the warring powers.

Steadily, petroleum gained on coal as the use of the internal-combustion engine doubled and redoubled, and as houses by the millions converted from coal heat to the more easily controlled oil heat. By the time World War II was over, the industrial world was powered chiefly by petroleum.

Fortunately, new oilfields were discovered or the world would have consumed most of the available petroleum. As it turned out, the richest stores of petroleum on the planet are in the Middle East.

Nearly three fifths of all known petroleum reserves on Earth are to be found in the territory of the various Arabic-speaking countries. Kuwait, for instance, which is a small nation at the head of the Persian Gulf, with an area only three fourths that of Massachusetts and a population of about half a million, possesses about one fifth of all the known petroleum reserves in the world.

This means that something has come to pass that no one would have dreamed of in 1900. The Moslem world dominates the energy picture, and the great industrial regions—the United States, Western Europe, and Japan—are, to an increasing extent, in their power. Of the industrialized nations, only the Soviet Union, which consumes far less energy than the United States, can still get by on its domestic production.

The United States, requiring more petroleum each year, is importing more and more, which produces a continuing and growing adverse balance of payments that is upsetting the American economy.

Yet in a way, it is all temporary, for at the present rate (allowing for the fact that the energy demands of the world will surely continue to increase as more nations grow industrialized), the petroleum reserves of the world will be consumed in about thirty years. The discovery of still more oilfields cannot postpone that day more than another decade or two.

What then? Do we dismantle our industrial civilization? We cannot do so without catastrophe. We must, rather, find new sources of energy.

It may become necessary, for instance, to return to coal, which is far more abundant than petroleum. Even at present levels of energy use, the coal reserves of the world will last us for two or three centuries. Coal is dangerous to mine, difficult to transport, and inconvenient to use, however. With all economy of energy use and despite any new techniques for using coal more easily and more efficiently, what happens when Earth's easily available coal reserves are gone?

Something more is needed, some new revolution even greater than the petroleum revolution of the early twentieth century.

One such possible development has already begun as an outgrowth of increasing knowledge of the atom. The enormous reservoir of energy locked away in the nucleus of the uranium atom was trapped in 1945 in the form of the uranium-fission bomb. That, however, was the result of a deliberately arranged runaway explosion. Could the energy of uranium fission be released in a controlled manner and put safely to use?

Yes, of course. As long ago as December 2, 1942, the very first nuclear reactor had produced controlled energy—but only very little. Through clever engineering, far more efficient reactors for the controlled production of fission energy were produced, and in January 1954, a nuclear submarine, the U.S.S. *Nautilus*, was launched, a submarine powered by uranium fission reactors. Many such nuclear submarines have since been built by the United States and by the Soviet Union.

In June 1954, the Soviet Union completed a small nuclear station designed to use uranium fission power to produce heat that would turn turbines and generate electricity for civilian use. It was the first reactor of this sort in the world. In October 1956,

Great Britain put Calder Hall into action, the first full-scale civilian nuclear power plant. In 1958, the United States built its first civilian nuclear power plant at Shippingport, Pennsylvania.

By the early 1970s, about 3 percent of the energy the United States uses is produced by such power plants. Fissioning uranium, the second energy revolution of the century, is at about the place now that petroleum was in 1900. It is possible that by the end of the twentieth century, fission will be supplying half our needs.

Uranium fission has its limits, however. In the ordinary nuclear power plants, it is a particular type of uranium atom, uranium 235, that undergoes fission. Only 1 out of every 150 uranium atoms is uranium 235. As long as we must depend upon this rare variety, nuclear power plants will not last us for much longer than coal will.

It is possible, however, to surround the fuel core in which uranium 235 is undergoing fission, with a shell of ordinary uranium, or one of a similar metal, thorium. Subatomic particles emerging from the fissioning core will then convert the ordinary uranium, or the thorium, into new kinds of atoms that can be separated out and that can then be made to undergo fission in their turn.

In this way, the uranium 235, as it is consumed, forms new fissionable fuel in the shell in quantity greater than that which is used up in the core. Such a nuclear reactor is called a "breeder reactor" because it breeds new fuel.

As long ago as 1951 the United States built an experimental breeder reactor at Arco, Idaho, so we know that such a thing is possible. It is only necessary to solve the engineering details that will make a large one both functional and safe.

The Soviet Union announced the completion of its first large-scale breeder reactor in 1973. This is, presumably, now in operation. The United States is pushing for the construction of one by 1980.

The breeder reactor would make the entire supply of Earth's uranium and thorium available as nuclear fuels, and it is estimated that the total amount of energy made available to us in this way would be 500 times as much as through coal and oil. Breeder reactors could keep mankind going for many thousands of years.

Unfortunately, this cannot yet be considered a happy ending. Mankind's use of energy has always polluted the environment with heat, with smoke, and with chemicals. This is going on now at a much higher level than ever before. It is not that pollution is something new, but that it is something much more intense now. The level of air pollution is raising serious health problems in some of our large cities.

Nuclear fission reactors, whether ordinary or breeder in variety, do not produce smoke, or the ordinary chemicals of burning. They do produce somewhat more heat for the amount of electricity they deliver than coal and petroleum do, however.

Worse yet, nuclear fission reactors produce radioactive waste, and that is far more dangerous to life than anything present in smoke. This radioactive waste must be carefully stored away in some place and in some fashion that will not allow it to enter the environment. This is difficult enough now, but what will happen when breeder reactors are producing such wastes at ten times, or a hundred times, or a thousand times the present rate?

And still worse, there is always the chance (not a large one, to be sure) that a nuclear fission reactor may go out of control and spread deadly radioactivity over hundreds of square miles.

Fission power can only be considered another stopgap, then, until still a third energy revolution is produced that will supply mankind with copious energy and minimum pollution.

One possibility is nuclear fusion, the kind of energy produced when small atoms of hydrogen are built up or "fused" into larger atoms. Nuclear fusion produces far more energy than nuclear fission does. It is nuclear fusion that explodes Earth's hydrogen bombs, and it is nuclear fusion that powers the Sun and the other stars.

Nuclear fusion can be made to use, as its fuel, a variety of hydrogen called deuterium. There is enough deuterium in our oceans to supply mankind's energy needs for many millions of years, and deuterium can be far more easily obtained than coal, petroleum, or uranium can be. Furthermore, nuclear fusion reactors cannot possibly go out of control, since only a tiny bit of hydrogen undergoes fusion at any one time. In addition, the amount of radioactivity produced is small and can easily be controlled.

The catch is that although scientists in the United States, Great Britain, and the Soviet Union have all been trying to work out methods for igniting a nuclear fusion reaction and then keeping it under control, a quarter century of research has not yet produced the answer. Still, progress is being made (if slowly), and there is a chance that the answer may come before the twentieth century is over.

Meanwhile, other long-term natural sources of energy are being investigated. There are the tides, which draw on Earth's rotation for their energy; there is geothermal energy, which draws on the heat of Earth's interior; there is solar energy which makes direct use of sunlight. All of these energy sources will last as long as mankind probably will, if only we can develop the necessary engineering to put them to practical use.

So it follows that although two major energy revolutions have taken place in the twentieth century—petroleum and uranium fission—and have raised the world's levels of comfort and affluence to record heights, we are now in an energy crisis anyway, and only a third major energy revolution can save us.

6 · The Electronic Revolution

In 1900, electricity was already in common use for devices that seemed the marvel of the age. The electric telegraph had come into use in 1844, the transatlantic cable in 1866, the telephone in 1876, the electric light in 1879. In these and in all other electric devices, the working was brought about by an electric current that ran through wires and that was controlled by switches that closed or opened the connection between the wires.

The key to something more advanced came in 1883, when the American inventor Thomas Alva Edison, who had produced the first practical electric light, was trying to make the filaments of those lights last longer. He sealed a metal wire into an evacuated light bulb near the hot filament to see if that would help. It didn't, but Edison noticed that the electric current seemed to leap the vacuum gap between the hot filament and the cold wire. This "Edison effect" seemed mysterious at the time.

In the 1890s, however, it was discovered that atoms were made up of smaller particles; that in the outskirts of the atoms were tiny electrons that carried an electric charge. It began to be understood that an electric current through a wire was the consequence of moving electrons. The Edison effect took place because electrons could be made to emerge from a heated wire and leap across the vacuum to the cold wire.

In 1904, the English electrical engineer John Ambrose Fleming put the Edison effect to use. He surrounded a hot filament in an evacuated bulb with a cylindrical piece of metal he called a "plate."

He then attached the filament and the plate to an alternating current. In an alternating current, electrons first flow one way in the wires, then in the other, changing direction sixty times a second. This means that electrons first pull out of the plate and pile into the filament, then pull out of the filament and pile into the plate, back and forth sixty times a second.

When the electrons pile into the heated filament, the heat drives them out into the vacuum, and they make their way across to the plate, where there is an electron deficit. The current flows.

When the electrons move into the plate, there is no heat to drive them into the vacuum and there is no current flow across the gap. This means that although the direction of the current changes sixty times a second, the circuit is closed only when the electrons flow into the filament. The evacuated bulb, with its filament and plate, acts as a "rectifier," changing a two-way alternating current into a one-way direct current.

In 1907, the American inventor Lee De Forest inserted a second plate between the filament and the original plate. This second plate was perforated with holes through which the electrons could pass, and it is called the "grid."

The grid is hooked up to an electric current in such a way that it carries a permanent electron deficit. The presence of this deficit particularly close to the filament tends to suck electrons out of the filament and makes the current flow greater than it would be if the grid were absent.

Furthermore, the extent of the electron deficit can be made to change and to vary quite rapidly. If the grid is hooked up to an electric current that is varying in intensity very rapidly, the grid's electron deficit will change in intensity in exactly the same way. As the deficit increases and decreases, the electron flow from the filament increases and decreases.

A small change in the electron deficit in the grid will result in a similar but much larger change in the amount of electrons sucked out of the filament. A tiny, varying current attached to the grid will produce a much larger current, varying in precisely the same way, between the filament and plate. The filament-grid-plate combination therefore acts as an "amplifier."

When a device contains circuits that are controlled by electron flow across a vacuum instead of through a wire, it is said to be an "electronic" device. The electron flow can be controlled very rapidly and delicately by variations in electromagnetic fields, far more rapidly and delicately than wire connections can be opened and closed. As a result, electronic devices can do things ordinary electric devices cannot.

Consider, for instance, the discovery, in 1888, of radio waves (like light waves, but much, much longer) by the German physicist Heinrich R. Hertz. It seemed possible to use beams of radio waves to carry messages over long distances, so that there could be communication without wires. On December 12, 1901, the Italian electrical engineer Guglielmo Marconi managed to transmit a beam of radio waves across the Atlantic Ocean from England to Newfoundland.

At first the radio waves could only go on and off to send Morse code. However, the radio waves could set up tiny electric currents in the antennae receiving them. Sound could make the radio wave intensity vary. The tiny electric currents would change in intensity according to the sound wave variations, and if they could be changed back into sound waves, you could hear what was being said thousands of miles away.

The electric currents were too tiny for the purpose, however, unless they could be amplified. As it happened, the filament-grid-plate combinations in an evacuated bulb could do the trick. It was when these devices were used that radio became practical for the general public. That is why the evacuated bulbs came to be called "radio tubes" in the United States.

From about 1920 on, all kinds of electronic devices that depended on tiny fluctuating currents that could be greatly amplified and delicately controlled by radio tubes, which were manufactured in countless varieties.

One particular electronic device involves a stream of electrons passing through a large evacuated tube and striking a surface coated with fluorescent material that glows on impact. The stream of electrons is controlled so as to sweep across the surface in successive lines, covering the entire surface in a fraction of a second, then repeating.

The intensity of the electron beam is controlled by the amount of light falling on different parts of the rear inner surface of a camera, so that the beam paints a picture on the surface that is exactly like the picture the camera sees. This is "television," but it is only practical if the electric currents and electron streams involved are controlled and amplified by radio tubes.

We have had television since 1948, and this, like radio, sound amplifiers, modern phonographs, and many other devices, is possible only because we have electronic controls and not merely electric switches.

Yet there are drawbacks to the marvel of the radio tube. The radio tube must be bulky because it must enclose enough vacuum to insulate the filament from the grid and plate, so as to prevent a short circuit. Then, too, glass is brittle and can break or develop leaks that would spoil the vacuum. Finally, the tube doesn't work till an electric current heats the filament red hot. This means that every electronic device using radio tubes will work only after a "warm-up" period.

In the 1940s, however, scientists at the Bell Telephone Laboratories grew interested in substances such as silicon and germanium. These conducted electricity, but not as well as metals did, so they were called "semiconductors." They conducted, however, only if they contained small amounts of certain impurities.

Every germanium atom has four electrons in its outermost region. Where a whole mass of germanium atoms are present in a crystal, this is a stable arrangement, and the electrons do not move. Suppose small quantities of arsenic are present, however.

Each arsenic atom has five electrons on its outskirts. The fifth electron is free to move, and it is that bit of electron excess that makes germanium a semiconductor. This can be combined with germanium containing an impurity that produces an electron deficiency.

This kind of electron excess and electron deficiency can be used to set up rectifiers and amplifiers that work just the same as radio tubes with their electron excess in the filament and their electron deficiency in the grid and plate. Devices with solid semiconductors replacing evacuated tubes are called "solid-state electronic devices."

An amplifier made up of semiconductors is called a "transistor." The first one was developed at Bell Telephone Laboratories on December 23, 1947, by W. B. Shockley, W. H. Brattain, and J. Bardeen. The transistor, in doing the work of a radio tube, corrects every disadvantage possessed by the latter. It doesn't need a vacuum, so it can be very small and cheap, and there is no glass to break or leak. Then, too, it works at room temperature, so there is no warm-up period necessary, no heated filament to break. The transistor is far more rugged and can last far longer than a vacuum tube.

Of course, for transistors to work, they must be composed of very pure materials to which just the right amount of specific impurities are added. In 1952, the technique of zone refining was introduced by William G. Pfann, and in this way germanium of extreme purity could be produced. That was what was needed to make transistors and solid-state electronic devices practical.

At once, electronic devices, with transistors replacing radio tubes, could be made much smaller. As early as 1953, transistorized hearing aids could be made small enough to be fitted into the earpieces of spectacles and plugged into the ear canal. It was the first dramatic example of "miniaturization."

Radios became small enough to slip into a pocket. Pacemakers, producing accurately timed electrical pulses of fixed tininess, could become small enough and reliable enough to be implanted into the failing hearts of thousands who were in this way kept in normal health. Chips the size of postage stamps can be filled with thousands of circuit elements each so small as to be scarcely visible to the naked eye.

We can summarize the situation this way:

In 1900, electronic devices of any kind were unknown, but now, three quarters of a century later, we cannot do without them. Minus transistorized electronic control, scarcely a communicating device would work. Airplanes would be grounded, and our sophisticated war machine would grind to a helpless halt. The work of government and of business would be thrown into absolute confusion.

With these devices we have our radios, our record players, our television, and innumerable devices for education, health, business, convenience, and amusement—devices that were undreamed of in the youth of our older citizens.

NOTE: The tale of the transistor is told in greater detail in Chapter 12 entitled "Happy Birthday, Transistor."

7 · The Computer Revolution

Calculating devices of one sort or another are as old as arithmetic. As soon as primitive man got the notion of numbers, he used his ten fingers as devices for keeping track and for adding and subtracting.

Then came the abacus, which used pebbles or counters in rows of ten, like many sets of fingers. One row kept track of the units, the next of the tens, the next of the hundreds, and so on. Centuries later came the move to mechanize the process, to make as little of it as possible the result of human manipulation, and as much of it as possible automatic.

In 1642, the French mathematician Blaise Pascal invented an adding machine consisting of a set of wheels connected by gears. Each wheel was marked with the digits from 1 through 9, plus a final 0. When the first wheel was turned a complete revolution to the 0, the second wheel was engaged and moved forward to 1. After ten turns of the first wheel, the second wheel was nudged all the way around to 0, and the third wheel moved forward to 1.

By turning the wheels backward and forward the proper amounts, numbers could be added and subtracted and the results read off the device. In 1674, the German mathematician Gottfried von Leibnitz arranged wheels and gears in such a way that multiplication and division could be carried out also.

These machines were curiosities and didn't come into wide use, but then, in 1850, an American inventor, D. D. Parmalee, patented a machine in which the wheels weren't moved by hand, but by pushing down marked keys. That was the "cash register," which was *the* calculating machine of 1900, and for a number of years afterward, too.

But even the most advanced calculating machine in general use in 1900 was scarcely any advance at all over counting on your fingers. Such a machine did only the simplest arithmetical computations and it had to be supervised at every step.

Yet there were signs of something more than that. If one had a simple and repetitive task to do, could one "instruct" the machine once and for all and have the machine do it without furthur human supervision?

In 1900, something of this sort was only three years old. It was the pianola or "player piano," which reached the peak of its popularity in the 1920s. A roll of stiff paper was perforated in a careful pattern, and that pattern was the instruction. By the action of foot pedals, air could be blown through those perforations and activate a piano key. By carefully planning the location of the perforations, a piano could be made to play Beethoven's *Moonlight Sonata* without anyone's hands actually touching the keys.

Such punch-card techniques had first been invented in 1804, by a French weaver, Joseph Marie Jacquard. The presence or absence of holes in this place or that on a rectangular piece of cardboard depressed or raised the threads and created automatic designs in silk weaving.

In 1822, it had occurred to an English mathematician, Charles Babbage, to use punch cards to guide a calculating machine. By using a proper combination of holes, a sufficiently complex machine could be instructed (or "programmed") to do every kind of mathematical operation known to man. Babbage spent years trying to build a machine that was sufficiently complex for the purpose, but he could not. His theory was perfect, but the mechanical techniques of the nineteenth century were insufficiently sophisticated.

Simpler punch-card calculating machines were working by 1900, though, and they were improved over Babbage's attempt by using electricity to move the gears rather than mechanical pushes and pulls.

As the decades of the twentieth century passed, there was a constant and growing need for mechanical devices that could do calculations of a more and more complex kind in less and less time. As science advanced, complicated equations of all kinds had to be solved; and as social life grew more complex, more and more statistics had to be analyzed. In the 1930s, when the Social Security Act was passed, the American Government found it had to keep track of most of its citizens in great detail. Then, in the 1940s, when income taxes began to rise steeply, it found that these had to be kept track of in a massive way as well.

In 1925, the American electrical engineer Vannevar Bush constructed a machine capable of solving complicated "differential equations." It was a successful version of what Babbage had tried to do a century before, but it still worked with mechanical switches and wasn't fast enough.

In 1937, Howard H. Aiken, who was working at Harvard for his Ph.D., worked out plans for a complicated device that would solve differential equations by using electrical switches rather than mechanical ones. The device was completed in 1944 and was called "Mark I."

It was the first large-scale automatic calculating device—the first machine we could call a "computer." It weighed five tons and had over three thousand electrical relays and five hundred miles of wiring. A punched-paper tape controlled it as though it were a player piano, electrical contacts being made through the holes, and once the tape was fed into its vitals, it worked automatically. It could add and subtract 23-digit numbers in 0.3 second and multiply them in 6 seconds. Answers could be produced on punch cards or on electric typewriters run by the machine.

Yet that wasn't fast enough. Even as Mark I was being constructed, World War II was raging, and the need for faster computations was growing (such as quickly calculating where and when to fire an anti-aircraft shell at a fast-moving airplane, allowing for its speed, direction, the wind motion, and so on—a problem on which the American mathematician Norbert Wiener worked).

In place of the electric switch, there came the electronic switch. Instead of closing a contact by an electromagnetic pull, there came the *much faster* stopping and starting of a flow of electrons in a vacuum tube.

By 1946, a vacuum tube device was completed under the direction of John P. Eckert and John W. Mauchly at the University of Pennsylvania. It was called "Electronic Numerical Integrator and Computer" or, in abbreviated form, ENIAC. It was the first *electronic* computer. It contained 19,000 vacuum tubes and was much faster than Mark I, but it weighed fully 30 tons. (Startling as it was for its time, ENIAC was retired after only seven years—hopelessly outmoded.)

The Hungarian-American mathematician John von Neumann suggested methods whereby the computer could not only store the data fed into it and the results it obtained, but also even the operating instructions. It was then not necessary to start from scratch each time you wanted to get the computer to do some work: If you already had the general instructions stored in the machine, you merely introduced the necessary modifications.

Then, once the transistor became practical in the early 1950s, it became possible to replace the comparatively bulky vacuum tubes by the much tinier and much more rugged transistors. The result was that computers became much smaller without becoming less complicated. Where ENIAC took up 1,500 square feet of floor space, a modern electronic computer (in the field of computers, 1946 is not "modern") of equal complexity could be fitted into a space about equal to that taken up by an average refrigerator.

Punch cards vanished, too, for it was found that data and operations could be stored on magnetic tapes far more concisely, so that even a moderately sized computer can have an enormous memory. Instructions can be fed into the computer by a typewriter. Different keys or combinations of keys would take the place of the holes in a punch card. As time went on, various "computer languages" were devised, each one closer and closer to ordinary English. Now it is almost as though you can talk to a computer and have it talk back to you.

Computers came to work so rapidly that almost any reasonable problem could be answered in a matter of millionths of a second. Even if many problems were fed into it at the same time, the computer could solve one after the other, and even the last person in line would not be aware of having had to wait.

In 1965, therefore, the concept of "time sharing" was introduced. Many different people could have a typewriter connected to the computer, some at considerable distances. Each could use the computer freely, taking his turn, without ever being aware of delay and with the feeling that the computer was all his.

The usefulness of the new device was such that as early as 1948, two years after ENIAC, small electronic computers were being produced in quantity; within five years, 2,000 were in use; by 1961, the number was 10,000; by 1970, the number had passed the 100,000 mark—and still going up rapidly.

Computers have come to be essential for keeping government statistics in order in these days of income tax, Social Security, and welfare. Scientists have used them in solving problems in every field. Businesses use them to keep track of a thousand matters, from warehouse items to orders to billings. Computer memories are serving everyone now, even when they are scarcely aware of it—when they buy an airplane ticket, for instance.

In fact, it is amazing how much of the world would come to a halt if we suddenly woke up to find that every computer was out of action—considering that until after World War II, not one of them existed. Space exploration would stop; our military machine would grind to a halt; our government and major businesses would find themselves helpless.

And what of the future? Computers can play chess, compose music, translate languages, understand spoken words. Computers can do *anything* that can be broken down into a set of logical instructions for arithmetical manipulations.

Will computers ever "take over"? So far, there's no question of that. Until now, they are loyal servants of mankind, doing what they're told, and when they make a mistake, it is never the computer that errs, but the man who has given it its instructions.

And the annoyance of the mistakes are completely overwhelmed by the convenience of what computers make possible for us at every moment.

8 · The Communications Revolution

The great communications discoveries of the 1800s involved electricity along a wire. In 1844, the first telegraph was built to carry messages in codes of dots and dashes along wires. In 1866, cables carried those signals under the Atlantic Ocean from the United States to Europe. Ten years later, in 1876, the telephone carried the human voice across wires.

Always it was wires. In the year 1900, communications still depended upon them, and where the wires did not go, the electrical signals did not go either.

Yet something else was in the wind. In 1885, the German physicist Heinrich Hertz had discovered "Hertzian waves," which were like light waves, but much longer, and which could penetrate matter easily. By forming those Hertzian waves in pulses, beaming them over a distance, and receiving them, signals could be transmitted without wires.

The man who first demonstrated "wireless telegraphy" dramatically was the Italian engineer Guglielmo Marconi. In 1896, he sent a Hertzian wave signal over a distance of nine miles; in 1898, he sent one across the English Channel; and on December 12, 1901, he sent one across the Atlantic Ocean. It is December 12, 1901, when the twentieth century was but one year old, that is considered the date of the invention of "radio." And Hertzian waves are now called "radio waves."

At first, radio waves could only transmit dots and dashes, as the wave production was turned on and off. The American physicist Reginald Aubrey Fessenden developed a special generator that would keep the radio waves going constantly but would make them more and less intense in a pattern that could mimic sound waves. The radio waves matched the sound-wave pattern at the point of transmission and then produced the same sound-wave pattern at the point of reception.

On Christmas Eve 1906, music and speech came out of a radio receiver for the first time. The efficiency with which this could be done was enhanced as radio tubes were developed and added to the radio circuits. This made it possible to make radio reception more delicate and to amplify the sounds produced greatly. Earphones were no longer necessary, and the radio receiver produced audible sounds by the time of World War I.

During the war, the American electrical engineer Edwin Howard Armstrong developed a device for so adjusting the radio wave that it became possible to tune a radio to the reception of that particular wave by turning a single dial. It was no longer necessary to be an expert to carry on radio reception. Anyone could do it, which meant that as the 1920s opened, the radio could become a household fixture.

The 1930s marked the beginning of the golden age of radio, as the whole world was knit together by a network of insubstantial waves and as national leaders could, for the first time, reach the entire nation at a moment's notice. Franklin Delano Roosevelt rallied the United States to the battle against the Depression, while in Germany, Adolf Hitler used similar methods for his own darker purposes.

But radio could only transmit sound. What about sight?

Men learned to send pictures across wires, as earlier they had sent sound. A narrow beam of light passed through the picture on a photographic film to a phototube behind. The more transparent a particular spot on the film was, the more light reached the phototube, and the stronger the electric current produced. The beam of light scanned the photograph by moving across it along a line at the top, then along another line slightly below, then along another farther below, and so on. In this way an electric current was produced that had a strong-weak pattern that matched the light-dark pattern of the film. The current was received at a distant point, and its pattern was converted back into light-dark.

In this way, "wirephotos" were transmitted between London and Paris as early as 1907.

To make pictures that move, the photograph must be scanned very rapidly, and another slightly different photograph scanned just as rapidly, and so on. Each completed pattern shows one "still" scene of a movie or of real life. Separate scenes, each still, but shown in rapid succession, seem to move, and so we have television. The signals can be carried by radio waves instead of wires, so that all the techniques worked out for radio could be used for television as well.

A form of television was first demonstrated in 1926 by the Scottish inventor John Logie Baird. However, the first practical television camera was the "iconoscope," patented in 1938 by the Russian-born American inventor Vladimir Kosma Zworykin.

In the iconoscope, the rear of the camera is coated with tiny drops of a metal that emits electrons when light hits it. The quantity of electrons emitted depends on the brightness of the light. The electron-emission pattern follows the light-dark pattern of the scene viewed by the camera. The electron pattern is converted into a similar radio-wave pattern, which is carried to a receiver, where it is turned back into a light-dark pattern.

The television receiver is coated with a fluorescent substance. The beam of electrons shooting at it scans the screen and produces a light intensity that matches the electron-beam intensity. In one thirtieth of a second the entire receiver is scanned and the eye sees a complete scene; then another in the next thirtieth of a second; then another, and so on. The scenes, all put together, seem to move and are, of course, accompanied by sound.

Experimental television was first broadcast in the 1920s, but television became truly practical only in 1947. Since then, it has become the giant of communication and of entertainment. It has replaced radio and, to a large extent, the stage and movie house as well. It has forced a contraction of the newspaper industry by taking over some of the news function, and a contraction of the magazine industry as well by taking over much of the advertising.

The world is now accustomed to seeing events as they happen —even wars—and to seeing, as well as hearing, its leaders. In the United States, political campaigns without television are now unthinkable.

In the mid-1950s, two refinements went a great deal toward increasing the effectiveness of television. By the use of three types of fluorescent material, designed to react to electric beams in red, blue, and green, color television was introduced. And "video tape," a type of recording with certain similarities to the sound track on a movie film, made it possible to reproduce recorded programs or events with better quality than could be obtained even from motion-picture film.

Even with radio and television, however, people are dependent on large and expensive transmitting stations that can only transmit over one of a limited number of wavelength channels. There are therefore only a few things that can be done by radio and television, and personal use of communication is still confined to the telephone and is still tied to its wires.

In 1945, the science writer Arthur C. Clarke pointed out the possibility of communications satellites, hovering over the Earth 22,000 miles above the equator (where its revolution would match the speed of Earth's rotation). Three such satellites, evenly spaced, receiving and relaying messages, could make it possible to send such messages from any spot on Earth to any other, as easily as it could be sent from one part of town to another.

In 1960, the first communications satellite, the very primitive Echo I, was launched. More sophisticated satellites followed soon, and the first commercial satellite, Early Bird, was launched in 1965. It could handle 240 voice circuits or 1 TV channel. In 1971, Intelsat IV was launched and is in operation now with an average capacity for 5,000 voice circuits or 12 TV channels.

It is now common to witness live broadcasts from the other

side of the world with the words "via satellite" appearing on the screen.

The great ease of long-distance electronic communication made possible by satellites is still tied to radio waves and therefore confines us to a limited number of channels. In the 1960s, however, instruments called "lasers" were developed. These can produce light in even wavelengths all moving along in the same direction, as radio waves do. The laser light, however, is much smaller in wavelength than radio waves are, and there is room for many millions of channels.

If messages are beamed to and from communications satellites by laser, there would be room for so many channels that every person on Earth could have his own, just as he has his own telephone number. He could reach anyone else on Earth with no trouble, transmitting sight as well as sound.

The printed word in such a world could be transmitted as efficiently as the spoken word. Facsimile mail, newspapers, magazines, and books could be readily available at the press of a button—anywhere. They could be read off the screen or printed out.

As communication improves and becomes more intensive, mankind will find it less necessary to live together in huge clusters. In a world of laser-satellite communications, it will be information that will be transported from point to point, not human bodies. Information can be moved electronically at the speed of light, and with far less energy expenditure than is required for transporting mass.

With unlimited numbers of television channels, conferences can take place in the form of images. The actual bodies, belonging to those images, fed with facsimiles of any necessary documents, can be anywhere on Earth.

By eliminating much travel, the necessity of moving some human beings, either for business that can't be performed by image or for travel fun and pleasure, would be a lot easier. The energy saving in transporting information rather than mass would go a long way toward defusing the energy crisis.

Then, too, education can become more versatile and perhaps more effective if electronic communication can be used massively.

And as people grow to know each other better the world over, it is very likely that the kind of hatred that is bred out of lack of knowledge will decrease. The nations can co-operate more effectively, and some form of planetary attack on planetary problems can take place more successfully.

It is not unlikely that a good beginning toward laser-satellite communications may be made by the end of the century, provided mankind can retain its equilibrium and solve its present problems. If so, it will represent a remarkable advance, considering that at the beginning of the century, mankind was still tied to the wire.

9 · The Space Revolution

In 1900, there was no thought that anyone would ever leave the face of the Earth altogether and travel to other worlds, except in the minds of story tellers. For centuries, writers had turned out fantasies of travels to the Moon. The most famous of these, prior to 1900, had been Jules Verne's *From the Earth to the Moon*, published in 1865.

Actually, the only device that could possibly carry a man outside the atmosphere was a rocket that carried its own fuel and oxygen (or carried fuel that burned without added oxygen). Such rockets had been used as fireworks for centuries, and even in warfare during the early 1800s. (Our national anthem speaks of the "rocket's red glare" during the siege of Fort McHenry in 1814.)

Few people thought of rockets for space travel, however. Jules Verne, in his book, had his spaceship hurled to the Moon out of a huge cannon. This is not practical if human beings are to be on board.

As the twentieth century began, a scientist, for the first time, took up the matter of space travel seriously. He was Konstantin E. Tsiolkovsky, a Russian physicist, who, in 1903, published the first careful mathematical study of the workings of a rocket-powered spaceship.

An American physicist, Robert H. Goddard, did the same independently. In 1919, he published a small book about rocket ships that went over the same ground that Tsiolkovsky had covered. But then Goddard went farther. He did not involve himself with theory only. He went on to try to build rockets.

In 1923, he began to test rockets that were not powered by solid fuels like gunpowder, something all earlier rockets had used. He used gasoline and liquid oxygen. These liquids could be allowed to flow into a combustion chamber at a fixed rate, and the burning could take place under controlled conditions. The exhaust gases would be ejected downward through a narrow opening, and the rest of the rocket would rise upward in accordance with Newton's law of action and reaction.

On March 16, 1926, on a farm in Auburn, Massachusetts, Goddard was ready to fly his first liquid-fuel rocket. It was only 4 feet long and 6 inches thick. It rose 184 feet into the air and reached a speed of 60 miles an hour. No reporters were present. No one, perhaps not even Goddard, knew that the first beginning of the space revolution had taken place.

Goddard continued to experiment farther, and in July 1929 he sent up a rocket that carried a barometer and a thermometer. This was the first instrument-carrying rocket. Neighbors complained of the noise and danger of the rocket firings, however, and Goddard was forced to move to a desert region of New Mexico, where he could experiment in peace.

There he worked out the notion of multistage rockets, where large rockets carried small rockets aloft. The small rocket could then fire after it had gained considerable speed and in a place where the air was very thin. Greater heights could in this way be achieved than the same amount of fuel would achieve in a single rocket.

In the early 1930s, Goddard finally fired rockets that reached speeds faster than sound and that rose a mile and a half into the air. But he worked alone and received no recognition. The American Government was not interested.

In Germany, there was greater interest. In 1923, a book on

rocketry was published in German by Hermann Oberth, and it made enough converts to bring about the organization in 1927 of a "Society for Space Travel." Wernher von Braun joined the society when he was a teen-ager.

The German society experimented with rockets, following the same path that Goddard was taking in America. The difference was that when Adolf Hitler came to power in Germany in 1933, he became interested in rockets for possible war use and began pouring government money into the research.

In 1936, a secret experimental station was built on the Baltic seacoast, and by 1938, rockets had been built there that were capable of flying 11 miles. During World War II, these rockets were developed into missiles carrying explosives. In 1944, Wernher von Braun, who led the project, put them into action as the famous V-2 rockets.

In all, 4,300 V-2 rockets were fired during World War II, and of these 1,230 hit London, killing 2,511 and wounding 5,869. The rocket weapon had not come soon enough to win the war for Germany. It had also come too late to bring Goddard recognition, for he died on August 10, 1945.

The American and Soviet governments, watching the work of the V-2 rockets, now realized the importance of the technique. In the final stages of the war in Europe, both tried to get as many of the German rockets and rocket scientists as possible. The United States got most of both, including Wernher von Braun.

The two nations then worked hard to build rocket missiles. By the 1950s, the old V-2 was a piddling affair compared to the monsters that were coming into existence. Both the Soviet Union and the United States developed "intercontinental ballistic missiles" (ICBMs), which could travel thousands of miles and land accurately on target. Such a missile, with a thermonuclear warhead, could do unimaginable damage.

But rockets were being used for scientific research, too. The United States used captured V-2 missiles to carry instruments high above the Earth's surface. One reached a height of 114 miles, five times as high as any plane or balloon could reach.

In 1949, the United States put a small American rocket on top of a V-2. When the V-2 had reached its maximum height, the small rocket took off and reached a height of 240 miles.

Such high-flying rockets brought back useful information about the nature of the upper atmosphere. They recorded temperatures, densities, wind speeds, charged particles, and chemical composition of the upper atmosphere.

Such rockets stayed in the upper atmosphere only a short time, however. It would be more useful if they could be made to curve into a horizontal path and take up an orbit about the Earth. What was needed was a rocket that could reach a height of 100 miles or more while going at a speed of at least 5 miles a second, and that could have its direction accurately controlled.

A rocket in orbit would be an "artificial satellite," and experts felt that one could be launched successfully by the time of the "International Geophysical Year" (IGY) that was scheduled for 1957. During that year the Earth as a whole would be studied by modern scientific methods by the world's scientists. On July 29, 1955, the United States announced officially that it would launch a satellite during the IGY.

The Soviet Union had been experimenting with rockets also, however, and to the surprise of the world, it launched a satellite first. On October 4, 1957, almost exactly one hundred years after the birth of Tsiolkovsky, Sputnik I was put into orbit and the Space Age began.

The United States was soon launching satellites of its own. On January 31, 1958, the first successful American satellite, Explorer I, was placed in orbit. In the years that followed, many hundreds of satellites were launched by each nation.

The satellites turned out to have a great many practical uses. Some were "weather satellites," which took many thousands of photographs of the Earth, outlining its cloud cover and air currents. Some were "communications satellites," which could receive radio waves, amplify them, and relay them back to Earth, thus making it possible for television communication to span continents and oceans. Some were "navigation satellites," which could be positioned from the ground so that maps of many parts of the Earth could be made with greater accuracy than ever before. Some were satellites that studied the surface of the Earth, some studied the space near Earth, some studied neighboring worlds, and some studied the distant reaches of the cosmos. And some, of course, were spy satellites, serving military purposes.

Satellites were sent to the Moon. On September 12, 1959, the Soviet satellite Lunik II hit the Moon. It was the first man-made object to land on another world. A month later, Lunik III circled the Moon and took pictures of the hidden side, which no one had ever seen before.

The Moon was studied in greater and great detail until, by 1966, both Soviet and American rockets were making soft landings on the Moon and sending back pictures taken from the surface. Other rockets were passing near Venus and Mars and sending back data taken from near those planets.

But the most glamorous part of rocketry was the task of carrying human beings into space.

On April 12, 1961, the first man was placed into orbit around the Earth. This was the Soviet cosmonaut Yuri Gagarin, who made a single turn about the planet before landing safely. On February 20, 1962, the first American astronaut, John H. Glenn, was placed in orbit and traveled three times around the Earth before landing.

In the years that followed, Soviet cosmonauts and American astronauts spent longer and longer times in space—entire days, even weeks. They went up in ships capable of holding two men, then three. In 1965, both cosmonauts and astronauts left the ship, while encased in spacesuits, and took "spacewalks."

After that, the task of actually reaching the Moon was left to the Americans, as the Soviet space program moved in other directions.

In December 1968, an American vessel, Apollo 8, carrying three men, traveled to the Moon. It didn't land there but circled it ten times at a height of less than 70 miles, then brought its crew safely home.

Two other Apollo vessels went to the Moon and back, each coming closer than the one before. They practiced the approach to the Moon itself in a "lunar module," which left the mother ship and returned.

Finally, Apollo 11 was launched on July 16, 1969. It was the heaviest spaceship launched up to that time (3,300 tons), and at 4:20 P.M. on Sunday, July 20, 1969, its lunar module, carrying American astronauts Neil Armstrong and Edwin Aldrin, reached the surface of the Moon. Armstrong was the first man ever to set foot on a world other than the Earth and, as his foot touched the ground, he said, "This is a small step for a man, but a giant leap for mankind."

So the twentieth century, which had begun with a trip to the Moon seeming only a science fiction writer's dream, saw this become actuality before seven decades had passed. Nor was that all. Five more successful round-trips were made to the Moon. Rockets have been sent as far as the giant planet Jupiter, and have sent back useful information. Men have remained in the scientific satellite Skylab for months at a time.

The Space Revolution, having taken place, is continuing, with ends that cannot yet be foreseen.

10 · The Universe Revolution

Even as man's personal horizon stretched out to the Moon in the course of the twentieth century, the universe itself, as we know it, stretched outward far more, incredibly more.

In 1900, it was known that the stars we could see in the sky, together with many millions we could not see, formed a lens-shaped collection called the Galaxy. The best estimate of its size in the early years of the twentieth century, was that it was 23,000 light-years across. (One light-year is equal to 5.9 trillion miles.) The Sun was thought to be located near the center of the Galaxy, and it was strongly suspected that nothing existed beyond the Galaxy.

In 1914, the American astronomer Harlow Shapley made use of a new way of estimating distance that involved certain pulsating stars called "Cepheids." He showed that parts of the Galaxy were hidden by dark clouds of dust and gas. Allowing for this, it has turned out that the Galaxy is 100,000 light-years across, with the Sun far toward one end of it. Two little foggy patches in the sky, the "Magellanic clouds," are collections of stars lying outside the Galaxy, about 150,000 light-years from our Sun.

What's more, by 1923, Cepheid stars were located in a foggy patch of light in the constellation Andromeda. That patch was hundreds of thousands of light-years away. To appear as bright as it did meant that it was a collection of stars as large as our own Galaxy. Looking about the sky, astronomers now realized that there were billions of galaxies like our own at immense distances.

One thing that could be learned by studying the light of these distant galaxies was whether they were approaching us or receding from us and, in either case, at what speed. In the 1920s, the American astronomer Milton La Salle Humason studied as many galaxies as he could and found that, except for one or two of the nearest, all were receding from us. The fainter they were, the faster they seemed to be receding. This fit in with the theory of an "expanding universe" in which the distances between all the galaxies were getting greater and greater with time.

The American astronomer Edwin Powell Hubble worked out a rule, in 1929, for estimating the distance of a galaxy from the speed at which it receded from us. This rule has been improved since. It now appears that the Andromeda galaxy, which is the nearest large galaxy outside our own, is about 2,300,000 light-years away. The most distant galaxies that can be made out in our largest telescopes are hundreds of millions of light-years away.

Then, in 1931, an American radio engineer, Karl Jansky, discovered that radio waves could be detected from various places in the sky. During World War II, radar was developed, and the devices used in radar turned out to be just right for studying the radio waves from the sky. Huge "radio telescopes" were therefore built in the 1950s.

One discovery made then was that certain dim stars gave off radio waves. Astronomers began to study those otherwise apparently ordinary stars and found their light hard to analyze. In 1963, though, the Dutch-American astronomer Maarten Schmidt demonstrated that these stars seemed to be receding at unheard-of speeds and were therefore immensely far away. They were not stars, but "quasistellar" ("starlike") in appearances. This word was shortened to "quasar."

Even the nearest quasar turned out to be over a billion light-years away. They are only visible because they are a hundred times as luminous as ordinary galaxies, but exactly what they are, no one knows.

In the past ten years, dozens of quasars have been found. In 1973, some were found that seem to be over 12 billion light-years away from us.

These very distant quasars are receding from us at over nine tenths the speed of light. Anything a little farther away would be receding at the full speed of light, and their light would therefore never get to us. The most distant quasars we can now see are, for that reason, nearly at the limit of what we are ever likely to see.

In 1900, then, we knew of only one collection of stars, our own, and thought it was only 23,000 light-years across. Now we have penetrated the universe to its observable limit, and know that there are many billions of star collections like our own, together with dozens of mysterious objects called quasars, which are up to 12 billion light-years away.

Our Galaxy, which is all we knew in 1900, makes up less than a millionth of a billion of the universe that we now see, study, and try to understand—less than three quarters of a century later.

11 · The Health Revolution

By 1900, the medical profession had already made some enormous strides. In 1796, the English physician Edward Jenner worked out the technique of vaccination, and where that was used, the dread disease of smallpox virtually vanished. It was the first infectious disease to be conquered by man.

In the 1860s, the French chemist Louis Pasteur advanced the germ theory of disease. In the wake of that, physicians learned how to inoculate people against certain diseases such as diphtheria and typhoid fever. A better understanding and practice of hygiene, greater care in water supply, and the use of quarantine contributed to cutting down the incidence of disease.

As the twentieth century opened, however, infectious disease was still the prime killer of man. In 1918, there was a worldwide influenza epidemic that killed more human beings in a few months than had been killed by World War I in four years.

The trouble was that though the various techniques worked out by nineteenth-century medicine helped prevent disease; there was little physicians could do once a disease took hold. (Yet nineteenth-century medicine had its points. As late as 1955, such techniques, developed by the American physician Jonas Edward Salk, put a virtual end to the terror of poliomyelitis.)

The twentieth century saw a search for any chemical that would attack some specific disease-producing bacterium within the human body and yet leave the body itself unaffected. The first seeker after such a chemical, the German physician Paul Ehrlich, referred to it as a "magic bullet." By dint of searching through hundreds of chemicals, Ehrlich, in 1909, found "arsphenamine" (No. 606 in his list), which worked, to a degree, against syphilis.

More than twenty more years passed, however, and other magic bullets were not found. Then, in 1932, a German biochemist, Gerhard Domagk, who was working with dyes, found "Prontosil," which, when injected into mice, seemed to have a powerful effect on any streptococci growing within them. Streptococci are bacteria that are a common source of infection in humans, and Domagk wondered if he dared try his dye on human beings.

The decision came about most dramatically. In 1935, his young daughter, Hildegarde, pricked herself with a needle; the insignificant wound was infected by streptococci, and soon the girl was suffering from a raging fever and was clearly dying. Nothing would help her, and the desperate Domagk tried the injection of large quantities of Prontosil. At once, Hildegarde began to recover.

It was soon found that only part of the Prontosil molecule did the work, and that part was a chemical known as "sulfanilamide." It was the first "wonder drug."

Sulfanilamide is a synthetic compound; it does not occur in nature. In 1939, however, the French-American microbiologist René Jules Dubos discovered there were compounds produced by bacteria that could be used to stop the growth of other bacteria. This turned the attention of scientists to the substances formed by microscopic organisms.

Back in 1928, a Scottish bacteriologist, Alexander Fleming, had reported that when cells of a certain mold, *Penicillium notatum*, got into a bacterial colony, they killed all the bacteria around them. No one had paid any attention at first, but after Dubos' discovery, scientists turned back to Fleming's report—especially since World War II was beginning, and it was expected that there would be an unusually high number of infected wounds to deal with soon.

The Australian-English physician Howard Walter Florey led the research, and the bacteria-stopping chemical produced by the mold was isolated and was named "penicillin." Penicillin was used, very successfully, on war casualties in Tunisia and Sicily in 1943. By the end of the war, the chemical structure of penicillin was worked out, and it was released for civilian use.

Because penicillin interfered with the life processes of various kinds of bacteria, it was called an "antibiotic" ("against life"). Though penicillin was the first, other antibiotics were soon discovered. The result of the discovery of antibiotics was that the world of medicine finally had its magic bullets, and infectious diseases declined in importance. All over the world, death rates dropped.

World War II produced another great victory for medicine, of a more indirect kind. In the 1930s, a Swiss chemist, Paul Müller, was looking for a kind of magic bullet against noxious insects. He wanted something cheap, stable, and odorless that would kill insects quickly and leave other forms of life unaffected. In September 1939, just as World War II was beginning, he found what he wanted in "dichloro-diphenyl-trichloroethane," a long name that was quickly abbreviated to DDT.

Switzerland put it to use at once in fighting the Colorado potato beetle. In 1942, DDT began to be produced commercially in the United States, and within the year it had its chance to prove its worth in connection with disease.

The Anglo-American forces had taken Naples in late 1943, and had found a typhus epidemic raging there. During World War I, typhus had been more dangerous than guns, and whole armies had been destroyed by it. Since then, though, it was discovered that typhus was spread by the body louse, and now there was a weapon against it.

Out came the DDT, and in January 1944, the population of Naples was sprayed. The lice died, and for the first time in history, a winter epidemic of typhus was stopped in its tracks. A similar epidemic was stopped in Japan in late 1945 after the American occupation.

The defeat in infectious diseases had left noninfectious diseases relatively more important, but there were advances here, too.

In 1900, several diseases were known that did not seem to be contagious and that might be cured by diet. Of these, beriberi was one. At least the Japanese Navy had put an end to beriberi aboard ship by changing the sailors' diet and adding barley, for instance.

A Dutch physician, Christiaan Eijkman, was sure beriberi was actually a germ disease. He studied it in Java, where it was common, and discovered (accidentally) in 1896 that when he fed chickens on the good, white rice the patients ate, the chickens fell sick with what looked like beriberi. When he fed them cheap, brown rice, they recovered. Eijkman convinced himself that beriberi wasn't a germ disease after all. There was something in the rice hulls that was necessary to life, the absence of which caused beriberi.

Various biochemists began to look for other such substances needed by the body in small quantities. In 1912, the Polish biochemist Casimir Funk, who had detected what chemists called an "amine group" in the rice hull substance, suggested the name "vitamin" ("life amine") for such substances.

Not only was beriberi beaten once the nature of vitamins and their importance in the diet was discovered, but also such diseases as scurvy, rickets, and pellagra. A disease called "pernicious anemia" was cured in 1926 when an American physician, George Richards Minot, began feeding patients liver—the liver contained a necessary vitamin.

A more balanced diet, with proper vitamin content, has produced men and women who are, in general, taller, stronger, and healthier than their parents and grandparents were.

Another important discovery had been made in 1902 by two English physiologists, Ernest Henry Starling and William Maddock Bayliss. They found that certain organs secreted substances into the bloodstream, substances that then roused other organs to activity. Starling called such substances "hormones," from Greek words meaning "rouse to activity."

It was by means of hormones that the serious and common disease "diabetes" was treated. It was discovered that the removal of the pancreas from experimental animals produced diabetes.

The pancreas, it was felt, produced a hormone needed to combat the condition, but no one could isolate it. Apparently, the hormone was a protein, and the pancreas contained digestive enzymes that destroyed the protein even while it was being isolated.

In 1922, two young Canadian physiologists, Frederick Grant Banting and Charles Herbert Best, discovered that if the pancreas duct in a living animal was tied off, those parts of the pancreas that produced the digestive enzymes withered away. What was left were the "islands of Langerhans" (portions of the pancreas named for the discoverer), which produced the hormone. The hormone could now be isolated and was named "insulin," from the Latin word for "island."

Other diseases came to be better understood and better treated as a result of hormone work, though in no case were the results as spectacular as in the case of diabetes, where sufferers could live out normal lives at the cost of periodic injections of the hormone. (Minot, who helped cure pernicious anemia, was himself a diabetic kept alive by insulin.)

As a result of the discovery of antibiotics, insecticides, vitamins, and hormones, to say nothing of improved surgical techniques and the use of X rays, electrocardiograms, electroencephalograms, electrophoresis, and other advanced biophysical and biochemical techniques for diagnosis and treatment, possibilities undreamed of in 1900 have become common, everyday occurrences.

To indicate the most dramatic change of all, the life expectancy of Americans has jumped from about forty in 1900 to about seventy nowadays. Every one of us has, on the average, the gift of thirty additional years of life as the result, almost entirely, of advances in medicine.

Victory is not total. There are diseases that remain unconquered, notably cancer and heart disease. Nevertheless, even with such deadly dangers uncontrolled, the life expectancy has been so extended, and so many people live to an advanced age that, for the first time in human history, old age itself has become an important disease to be studied.

Every year, the science of gerontology (the study of the phenomenon of old age) is becoming more important as a medical specialty, and every year physicians are zeroing in further on the possibility of making old age a stronger, healthier, more comfortable, and more prolonged time of life for all of us.

12 · Happy Birthday, Transistor

On December 23, 1972, we celebrated the silver anniversary of the invention of a marvelous little gadget. The accent here is on the "little," because what it gave us was not something really new, but something unprecedentedly tiny.

The device, called the transistor, made it possible for us to take advantage of the marvels of technology that we were beginning to enjoy, without having to devote large volumes of space to them. Electronic devices became smaller and less energy-consuming, small enough to be fired into space, tiny enough to be buried in a person's heart. The advance made through the transistor pointed the way toward ever more compact, ever more precise, and ever more versatile computers that have as their ultimate goal something as compact, precise, and versatile as the human brain—or beyond.

And all in twenty-five years!*

But science is all of a piece, and there are no real beginnings or endings. To understand the transistor, we have to go back well

---

* This essay was originally published on the twenty-fifth anniversary of the invention of the transistor.

beyond its official birthday to 1883, when the American inventor Thomas Alva Edison was striving to improve his recently invented electric light.

The greatest flaw in his invention was the way in which the glowing filament, in its glass-enclosed vacuum, slowly evaporated as it remained at white-hot temperature, and thus thinned and broke with continuing use. Edison, in his usual manner, tried everything he could think of to slow down this evaporation and lengthen the life of his light bulb.

One of the things he tried was to seal a metal wire into a light bulb near the hot filament. Perhaps the presence of the wire would somehow lengthen the life of the filament. It didn't, but in testing the matter, Edison noted something odd. A distinct electric current could be detected in the metal wire, even though it was not connected to any source of electricity.

There was, to be sure, electric current in the filament, but that was separated from the wire by a vacuum gap. Could it be that the current had leaped the gap? Edison reported this "Edison effect" (as it came to be called), and took out a patent on it. Since he could find no practical use for it, he let it slide.

In the next decade, however, a particle that was much smaller than an atom was identified, the first of the now-numerous "subatomic particles" to be studied. It was called the "electron."

All atoms, it was quickly found, contained electrons as part of their structure, and every electron carried a fixed quantity of something called an "electric charge." An atom contained electrons in its outer regions, and in its interior possesses other particles, called "protons," each of which carries an electric charge equal in quantity to that on an electron but of an opposite nature. Purely arbitrarily, we call the electron's charge "negative" and the proton's charge "positive." An ordinary atom contains equal quantities of negative and positive charge, so that as a whole it is electrically neutral.

As it happens, opposite electric charges attract, and similar charges repel. An electron and a proton attract each other, while an electron repels another electron and a proton repels another proton.

The negatively charged electron, on the outskirts of the atom, is far lighter and more mobile than the positively charged proton at the atomic center. While the protons stay put, then, the electrons can move freely under the proper circumstances. It was found that the electric current was associated with a movement of electrons.

Once this connection of electrons and electricity was understood, the Edison effect quickly ceased to be a mystery. In experiments conducted between 1900 and 1903, the British physicist Owen Willans Richardson showed that when metal filaments were heated in a vacuum, electrons boiled out of the filament's atoms in a kind of subatomic evaporation. This explained the Edison effect. An electric current, in the form of speeding electrons, *did* leap the vacuum gap from white-hot filament to wire. (For this, Richardson eventually received the Nobel Prize for physics in 1928.)

In 1904, the English electrical engineer John Ambrose Fleming put the Edison effect to brilliant use. He surrounded the filament in an evacuated bulb with a cylindrical piece of metal called a "plate."

Imagine the filament and plate hooked up to a source of alternating current. The current will run from the filament through the outside circuit to the plate, then it will reverse and run in the opposite direction, from plate to filament, then reverse again, and so on. The alternation in direction takes place many times a second. The thin wire of the filament offers sufficient resistance to the electric current so as to heat up through a kind of electrical friction; the broader plate offers much less resistance and does not heat up.

When the current moves in one direction, electrons suck out of the plate, leaving it with a net positive charge, and pile into the filament, which gains net negative charge. When the current moves in the other direction, the situation is reversed and the plate gains the negative charge, while the filament has the positive charge.

Many times a second, then, the situation reverses itself: filament negative and plate positive; then filament positive and plate negative; and so on.

When the hot filament is negatively charged, it has an excess of the negatively charged electrons, and can more easily emit them. It can all the more easily emit them since the plate is at that time positively charged, and the positive charge attracts the electrons and helps suck them out of the filament. Between the push of the filament and the pull of the plate, the electrons leap the gap in quantity, and a current races through the entire circuit.

When the filament is positively charged, there are few electrons available for emission. What's more, the plate is at that time negatively charged, and the negative charge repels any electrons that may show up. Nor can the cold plate emit any electrons of its own. As a result, electrons do *not* leap the gap. There is what amounts to a break in the circuit, and no current flows.

In other words, although the outside source of current is alternating, the actual current that makes its way through the evacuated bulb is direct, and flows in one direction only—from filament to plate—and never in the other. The evacuated bulb, converting two-way motion to one-way, is a "rectifier."

Fleming called his device a "valve" because it served to turn the electric current on and off, as an ordinary mechanical valve serves to turn the flow of water on and off.

Until Fleming's time, electric current had been turned on and off by the mechanical opening and closing of switches, and devices of this sort are "electrical" in nature. Fleming's valve controlled the flow of electricity by attracting or repelling electrons, and could work far more delicately and rapidly than any mechanical switch. Devices making use of such control of electron flow are "electronic" in nature.

In 1907, the American inventor Lee De Forest carried electronic control a step farther. He inserted a second plate between the filament and the original plate. This second plate was perforated with holes and was called a "grid."

The grid can have a positive charge imposed upon it. Being closer to the filament than the plate is, the grid's positive charge is particularly effective in sucking negatively charged electrons out of the filament and sending them on through its own holes to the plate to complete the circuit and keep the current flowing.

The grid can in this way cause a larger current to flow in its presence than would flow in its absence.

Furthermore, suppose the positive charge on the grid can be made to waver in amount. When the positive charge on the grid rises even slightly, the electrons come out of the filament in considerably greater quantity; while if it falls even slightly, the electrons come out in considerably lesser quantity. A small rise and fall in the charge on the grid will impose a correspondingly large rise and fall in the current passing from filament to plate, and the pattern of rise and fall in the large current will exactly mimic the pattern of wavering of positive charge on the plate. The "valve" is now more than a valve; it is an "amplifier."

Radio waves are capable of being converted into tiny electric currents, which can be made to affect the grid. As the amplitude of the radio waves rises and falls in a fashion that matches the sound wave imposed upon it, the grid's positive charge rises and falls in exact correspondence. The radio wave energy is low and the fluctuation of charge on the grid is tiny. Were that all, the sound emitted by radios would be too feeble to be generally useful in the home.

But that fluctuating charge on the grid controls a much larger current passing through the evacuated bulb, and it was such amplifiers that made the radio a means for entertaining (and once in a while instructing) the average American home. As a result, the amplifiers came to be known as "radio tubes" in the United States.

However, it is not only in radios that radio tubes were useful. They were all that served (prior to December 23, 1947) to make any electronic device practical—record players, public-address systems, and so on.

It seems rather ungrateful to complain of the radio tube when its services to technology have been so enormous, but it did have certain disadvantages, which we can see particularly clearly in retrospect.

Each radio tube had to be fairly large, since enough vacuum had to be enclosed for filament, grid, and plate to be far enough apart so that electrons wouldn't jump the gap until encouraged to do so. This meant that radio tubes were relatively expensive, since they had to be manufactured out of considerable material. Furthermore, since radio tubes were large, any device using them had to be bulky and could not be made smaller than the tubes they contained. Then, too, as devices grew more sophisticated, more and more tubes (each designed to fulfill a special purpose) were required, and bulkiness became more pronounced. Computers that made use of thousands of radio tubes were simply enormous.

Other difficulties were that radio tubes were fragile, since glass is brittle. They were also short-lived, since even the tiniest leak eventually ruins the vacuum, and since filaments will finally evaporate and break when under constant and repeated high temperatures. They were energy-consuming, since the tubes won't work unless the filament is maintained at a temperature high enough to cause it to emit electrons. Finally, they are time-consuming, since the tube won't start to work until the filament is first warmed up to the necessary high temperature.

(Anyone past the first flush of youth remembers the irritating "warmup" that elapsed after the switch-on before the radio started emitting sound, and the frequency with which a repairman had to replace this tube or that.)

The correction of every single one of the deficiencies of the radio tube came about because scientists at the Bell Telephone Laboratories grew interested in substances such as silicon and germanium in the 1940s.

Silicon and germanium are elements that have some of the properties we associate with metals. They are sometimes called "semimetals" in consequence. In particular, they can conduct electricity—not nearly as well as metals, to be sure, but far better than such typical insulators as glass, silk, and rubber. So silicon and germanium are examples of "semiconductors."

It was the semiconducting property that particularly caught the eye of the Bell scientists, particularly since that seemed to depend very strongly on the nature and quantity of the impurities in the substance, rather than on the substance itself.

Absolutely pure germanium, for instance (and whatever I say henceforth for germanium applies to silicon, too, be it understood), does not carry a current at all well and would not, in itself, be considered a semiconductor.

To see this, consider that each germanium atom has four outer electrons, at least some of which would have to move, somehow, if germanium were to conduct a current.

Each germanium atom will, however, share each one of its four outer electrons with a neighboring germanium atom. The result is that in a germanium crystal (where germanium atoms are arranged in a very regular pattern), every germanium atom is surrounded by four other germanium atoms. Between each pair of neighboring germanium atoms are a pair of electrons, one contributed by each of the neighbors. Around any individual germanium atom, then, are eight electrons, four of which are its own and four of which are contributed, one each, by the four neighbors.

All this is a very stable arrangement. It takes a great deal of energy to disrupt this electron pattern. With these electrons motionless, pure germanium will not conduct a current at all well.

But suppose a tiny bit of arsenic is added to the germanium crystal and is distributed evenly throughout. Every once in a long interval, an arsenic atom will be found in place of a germanium atom somewhere in the even atomic pattern that builds up the crystal.

And, as it happens, each arsenic atom has, not four, but *five* outer electrons.

Four of the outer electrons of the arsenic atom can play the role of germanium's four outer electrons. The arsenic atom, too, will be surrounded by eight electrons, four of its own and one each from its four neighbors. That leaves the fifth arsenic electron homeless. The eight electrons around each atom, whether arsenic or germanium, fill the available space, and that fifth electron of each occasional arsenic atom can move about easily, being held nowhere.

In the case of any electric current, the negatively charged electrons are repelled from the negative end of the circuit (the negative electrode) and are attracted toward the positive end of the circuit (the positive electrode). If an arsenic-containing germanium crystal is made part of the electric circuit, those loose fifth electrons therefore move away from the negative electrode and toward the positive electrode. Current flows through the germanium crystal easily enough to make it a semiconductor.

But suppose that to the original totally pure germanium crystal, a trace of boron is added, rather than arsenic. In that case, here and there in the germanium crystal, there will be a boron atom instead of a germanium atom—and the boron atom has only three outer electrons.

The boron atom in the germanium crystal is therefore surrounded by only seven electrons, three of its own and one each from its four neighbors. There is room for eight electrons surrounding the atom, and the place where the eighth electron ought to be remains unfilled. It represents a "hole" in the electron pattern.

Imagine such a germanium crystal made part of an electric circuit. Under the push of the electric current there is a tendency for the electrons to move toward the positive electrode. Most of the electrons can't respond. Not only are they held tightly in the eight-electron pattern, but also, even if one moved, there would be no place for it to move to, since all the places where an electron might go are already filled.

Except (if boron atoms are present) where an electron happens to be adjacent to the hole around one of the boron atoms. An electron from the negative-electrode side of the hole, moving toward the positive electrode, could slip into the hole. Now there is a new hole where the electron that had moved used to be. The new hole is slightly nearer the negative electrode. Another electron slips into it, and the hole is displaced to a position slightly nearer still to the negative electrode.

In a steady ticking, electrons sidle toward the positive electrode, each one moving into a hole, while the holes themselves move steadily toward the negative electrode.

The holes in the boron-containing germanium crystals behave like positively charged particles moving in the opposite direction to that in which the electrons move. The holes move toward the negative electrode in the same way and at the same speed that electrons move toward the positive electrode. The boron-containing germanium crystal will conduct an electric current exactly as an arsenic-containing germanium crystal will. Both are semiconductors.

So there are two types of semiconducting germanium crystals. There is the "n-type" (for "negative"), where there are negatively charged electrons moving toward the positive electrode; and there is the "p-type" (for "positive"), in which there are holes, behaving as though they were positively charged particles, moving toward the negative electrode.

Suppose that these two types of semiconductors are used in combination. Imagine a germanium crystal containing a trace of arsenic on one side and a trace of boron on the other; the first end n-type, and the other p-type. Suppose also that such a germanium crystal is made part of a circuit in which the negative electrode is attached to the n-type end, and the positive electrode to the p-type end.

The loose electrons in the n-type end migrate out of that end and move across the p-type end toward the positive electrode. The holes in the p-type end migrate out of that end and move across the n-type end to the negative electrode. Neither electrons nor holes get used up. From the negative electrode more electrons enter the n-type end. Electrons from the p-type end leave the crystal in their motion toward the positive electrode, leaving more holes behind. The net result is that when the circuit is hooked up in this fashion, electric current flows across the germanium crystal.

But suppose the electric circuit is hooked up in the other fashion, with the positive electrode attached to the n-type end, and the negative electrode to the p-type end. Now the electrons in the n-type end are attracted toward the positive electrode in their immediate neighborhood and pull away from the center of the crystal. The holes in the p-type end are attracted to the negative electrode in their immediate neighborhood and also pull away from the center of the crystal. The center of the crystal is left with neither excess electrons nor holes, but with atoms surrounded by an even eight electrons apiece. Such eight-electron atoms are nonconductors, and there is *no* electric current through the crystal under such conditions.

But what if an alternating current is attached to the two sides of the crystal? In that case each side is attached to a negative electrode, a positive electrode, back to a negative, then positive, in rapid alteration. Whenever the negative electrode is at the n-type end and the positive electrode at the p-type end, the current flows across the crystal; whenever the arrangement is reversed, it does not flow. Despite the fact that such a crystal is attached to an alternating current source, only a direct current gets through. The crystal acts as a rectifier as Fleming's valve does, the n-type end and p-type end acting as filament and plate, respectively.

Whereas Fleming's valve works because of the vacuum it contains, the crystal is a solid object from end to end. A germanium crystal that does the work of a vacuum tube is therefore called a "solid-state device."

It is possible to prepare a particularly useful solid-state device by using a germanium crystal in which both ends are n type, with a thin p-type center separating them. The p-type center acts as the grid does in a radio tube. If a varying current source is attached to the p-type center, it can control a similar and much larger current passing through the crystal—which then acts as an amplifier.

The men at Bell Laboratories who developed this solid-state amplifier and who completed the task on December 23, 1947, were William Bradford Shockley, Walter Houser Brattain, and John Bardeen. Their coworker John Robinson Pierce, suggested that such amplifying crystals be called "transistors" because they *trans*ferred an electrical signal over what would ordinarily be a re*sistor*.

The transistor, in doing the work of a radio tube, corrected every disadvantage possessed by the latter. In the first place, the transistor requires no vacuum to insulate the various elements of its structure from each other; portions of the crystal can themselves act as insulators. At once this wiped out the need for expensive vacuum technology or for fragile glass enclosures.

The crystals are far more effective as insulators than a vacuum is, so that the elements of the transistor can be separated by microscopic distances and still be insulated. The transistor can therefore be tiny in size—and cheap.

The transistor works at room temperature, so there is no warmup period necessary. Since nothing has to be heated and kept hot, there is no large energy consumption. Since there is no delicate filament to evaporate and break under continuing use, the transistor is far more rugged and can last far longer than a vacuum tube.

To be sure, the transistor did have one disadvantage to begin with. In order for it to work, it had to be composed of a very pure material to which just the right amount of specific impurities were added. If an impurity is present in slightly too great or too small an amount, or if the wrong impurities are inadvertently present, the transistor would not work properly.

Efforts were made to devise techniques for purification that would meet the need. In 1952, for instance, the technique of zone refining was introduced by William G. Pfann. In this technique, the end of a rod of, let us say, germanium is placed in the hollow of a doughnut-shaped heating element. The end of the germanium is heated until it softens.

The germanium is then slowly moved through the doughnut hollow so that the softened portion progresses along the length of the rod. The impurities in the rod tend to remain in the softened zone and are literally washed to the other end of the rod. After a few passes of this sort, the other end of the rod is cut off, and what is left of the germanium rod is unprecedentedly pure.

By the middle 1950s, then, the transistor had become a thoroughly reliable device, and there were the beginnings of mass production. The clearly apparent usefulness of the device led to a Nobel Prize in physics for Shockley, Brattain, and Bardeen in 1956.

Of all the advantages of the transistor, the most important was its smallness. As early as 1953, transistorized hearing aids were small enough to be fitted into the earpieces of spectacles and plugged into the ear canal. It was the first dramatic example of what the transistor would mean to the cause of "miniaturization."

For twenty years now, the advanced devices of our technological world have been shrinking.

The bulky piece of furniture we called a radio (on four legs and almost large enough to serve as a china closet) became a small box we could slip into our pocket and run on the energy output of tiny batteries. A pacemaker producing accurately timed electrical pulses of fixed tininess could become small enough and reliable enough to be implanted into the failing hearts of thousands, who were in this way kept in normal health.

There are computers which, if they depended on mechanical gears alone, would fill the Pentagon and take years to answer the complicated questions we are now used to asking. With radio tubes they would fill a large room and take minutes to answer. Transistorized, they stand in the corner and answer in fractions of a second.

And those simpler calculating devices that used to sit on desktops are now squeezed into small slabs that fit easily into the palm of the hand.

Without the transistor, there would be no space age. The small satellites we send out into space could do very little in the way of observing, recording, and sending back information were it not that tiny devices (tiny enough only by grace of transistors) could be packed into them.

Without the transistor, the necessary complexities required to design and build a rocket ship capable of sending men to the Moon and bringing them back alive, the calculations required to predict and control ignition, thrust, and trajectory, would demand computers too large, too slow, and too expensive to be practical.

Miniaturization continues. A chip the size of a postage stamp can be filled with thousands of circuit elements each so small as to be scarcely visible to the naked eye.

All of modern technology now depends upon these chips of semiconducting crystals. Without them, scarcely a communicating device would work. Airplanes would be grounded. Our sophisticated war machine would grind to a helpless halt. The work of government and of business would be thrown into absolute confusion.

It is hard to realize that it is only the silver anniversary of the birth of the transistor that we are now celebrating; harder to realize that the revolution it has brought about has become so extensive and intensive in a mere twenty-five years; and hardest of all to believe that it has all passed over our heads without our noticing, so that so few of the general public know that the anniversary has come and what it has all meant.

13 · The Whole Message

In 1971, the Nobel Prize for physics went to Dennis Gabor, a Hungarian-born British subject now working in America. He had earned the award in 1947, nearly a quarter century before, by inventing a process of recording images in a way that reproduced far more information than any other technique known. Because it contained virtually all the information, in fact, he named the process "holography," from Greek words meaning "the whole message."

For sixteen years, the process and the name slumbered in technical journals. Then, in 1963, two electrical engineers at the University of Michigan, Emmett N. Leith and Juris Upatnieks, carried the Gabor technique a step farther and made the front pages of newspapers.

Where Gabor had worked with electron waves and had applied his techniques to improving the images formed by electron microscopes, Leith and Upatnieks applied it to light. They produced a transparent sheet of film that was grayish in color, like an underexposed photographic film, and used it to produce a three-dimensional image in remarkably fine detail. And they did it without lenses.

How is it done?

To begin with, let us consider photography, a process that is by now quite familiar to us (though when it was first developed, over a century ago, it seemed just as mysterious to the public).

The foundation of photography is the ability of light to initiate certain chemical changes. Without going into detail at all, we can say that light can cause a colorless solution of a certain type to precipitate tiny black granules. If the solution is mixed with gelatin, coated on a film, and allowed to dry, the entire film will turn dark if exposed to light briefly and then treated with appropriate chemicals.

Suppose, however, that the film is exposed to light only indirectly, that it is allowed to shine on some object, and that only the portion of the light that is reflected in the proper direction shines on the film.

Some parts of the object will reflect light more efficiently than other parts; some parts of the object will reflect light directly toward the film, while other parts direct the light more or less away from the film; some parts will scatter the light that falls on them, sending it in many reflections, while others will reflect the light without scattering.

The result of all these differences in detail is this: The reflected beam of light will possess fine differences in brightness from point to point. When such a reflected beam of light enters our eyes, the pattern of varying light and dark is turned into a pattern of electrical impulses in the optic nerve. Our brain interprets the pattern in such a way as to give us an idea of the shape, the color, the texture, and so on, of the object that has reflected the light. We "see" the object.

But what if the same reflected beam of light falls on the photographic film? The pattern of varying brightnesses in the beam would then reproduce itself on the film. At a point on the film where there impinges a portion of the light beam that is quite bright, a considerable amount of chemical change is induced. Upon the proper treatment, that point becomes dark indeed. Where there impinges a dim portion of the light beam, there is little chemical change, and that part of the film remains light.

To produce a proper pattern on the film, it must be enclosed in a box in order to prevent light striking it from anything other than the object we want to record. Then, too, from each point on the object a sheaf of reflected light fans out. If all this light enters the opening in the box, each part of the film would be subject to some light from every part of the object, and the result would be a featureless blurring of the entire film. To prevent this, a lens is placed in the opening. Light passing through the lens is collected into a focus and brought to the film in an orderly fashion. (The light-recording part of the eye, the retina, is also enclosed in a box, the eyeball, and behind the eye's opening, the pupil, there is also a lens.)

The light focused on the film by the lens in the camera produces a reproduction of the object that reflected the light—but in reverse. The brighter portions of the beam are recorded as dark spots on the film, and the dim portions as light spots. The result is a "photographic negative."

If a featureless light is made to shine through a photographic negative so that it falls on a fresh film, the process is reversed again. All the dark places on the negative produce a dim portion of the beam passing through and are recorded as light places on the new film, and vice versa. The new result is a "photographic positive," which records the light and dark pattern exactly as it was reflected from the object.

If certain dyes are added to the film, advantage can be taken of the fact that some objects reflect light of particular wavelengths. By superimposing three images on the film, each involving a different wavelength of light, a photographic positive is produced that shows color rather than mere dark and light.

Assuming that the photography has been properly conducted, that the proper amount of light has entered the camera, and that the lens has been properly focused, then one "sees" the object on the film as in reality, and the image is recorded for as long as the film endures.

But is the image really *exactly* like the reality? Not really. Actually, the aspect of reality that the photographic image produces is quite incomplete.

Suppose you look at an object, say two chessmen on a chessboard, through a rectangular opening so as to make the object and its surroundings appear similar to the photographed image on the rectangular film. What then are the differences between real object and the image? (Of course, we can touch and feel the object and not the photographic image, but let us confine ourselves to visual properties only.)

Clearly, it is possible to tell image from reality by the sense of vision alone. Suppose we shift our head slightly as we look at the real object through the opening. What we see shifts, too. From one position, the nearer chessman may obscure the one behind it, but as we move our head, the farther chessman seems to move out somewhat from behind the closer. We see the real object in three dimensions and can look around an obstacle by moving our head.

This is not possible in the film. The film may give an illusion of three dimensions. A more distant object will look smaller than a similar object that is closer; the lines of a chessboard may show perspective. Still, however clever the appearance of three dimensions on a photographic image, the appearance remains an illusion and nothing more. No matter how we shift the position of our head, the image we see never changes; we see one view and one view only.

Another difference is this: In looking at a real object, we can focus our eyes on a nearer object, leaving a farther one out of focus; or, in reverse, focus on the farther at the expense of the nearer. We can move back and forth from one focus to the other at will. The image, on the other hand, has a single focus. If the farther chessman is photographed a bit out of focus so that the nearer one is clear and sharp (or vice versa), nothing we can do with our eyes can bring the out-of-focus portion into focus.

The reason for these limitations of the ordinary photograph is that we are recording the intersection of the light pattern with a flat, featureless surface (the photographic film). The intersection, not surprisingly, therefore has the properties of a flat surface, and in the process the reflected beam of light loses all its three-dimensional information. Photography ("the light message") is not holography ("the whole message").

But suppose we record the intersection of the light pattern with something more complex than a featureless flat surface. Suppose we record the intersection of the light pattern with *another* light pattern.

A beam of light consists of very tiny waves. The pattern in a beam exists because some light waves have a greater amplitude than others (moving farther up and down). This means the pattern is brighter in some places than in others. It also exists because some light waves are longer than others, which means the pattern shows different colors from place to place.

If two beams of light cross each other at an angle, particular waves in one beam may happen to match particular waves in the other. Both waves move up and down together. The result is that they reinforce each other. In combination, the two move up and down farther than either would separately. The combination of waves is brighter than either alone would be.

On the other hand, some waves in one beam may happen to meet waves in the other beam in reverse. One wave may be moving up as the other is moving down. In that case, the waves partly (or even entirely) cancel each other, and the combination of waves is dimmer than either alone would be.

In this way, when two patterns cross, the waves interfere with each other to form a new pattern of light and dark that wasn't present in either of the two patterns that had crossed. The new pattern is an "interference pattern."

If you have an interference pattern, you can, in theory, work out two patterns that could, in combination, form the interference pattern. The trouble is that any of an infinite number of combinations could have formed it, and there would be no way of deciding exactly which combinations did the job in reality.

Of course, if you knew one pattern of the two that formed the combination, you could calculate the other. To do it most easily, you would want the known pattern to be as uniform as possible. If one beam of light were simply uniform from one side to the other, with no variations in brightness or color, then from the interference pattern produced when it crosses another beam, the pattern of the other beam could be determined.

But how are we going to get blank light to act as a "reference beam"? Ordinary sunlight won't do. Sunlight might look blank and patternless, but it consists of a mixture of many colors, of light waves with a whole range of wavelengths. To work out the components of an interference pattern, where the simpler of the combining beams is as complicated as apparently featureless sunlight, is impractical.

How about producing a single color of light by heating some chemical substance that will then emit a single wavelength of light? Even that is not enough, for some of the light waves go in one direction, some in another. Even a beam of ordinary "monochromatic" ("one color") light is still not really featureless.

In fact, when Dennis Gabor first worked out the techniques of using interference patterns, there was no conceivable way they could be used for light waves. Nowhere, either in the universe or in the laboratory, did a beam of light exist in which all the waves were of exactly the same length and were moving in exactly the same direction. Unless such a beam could be found or could be made, there was no reference beam blank enough to allow us to calculate the pattern of the other beam with certainty from the interference pattern of the two. Gabor used his technique for wave forms other than light waves, where a calculation could be made.

But then, in 1960, the American physicist Theodore Howard Maiman constructed the first laser. The laser is a device that produces a powerful beam of light in which all the waves are of exactly the same length and in which all move in exactly the same direction. At last, the truly blank reference beam existed.

The laser beam contains no pattern, no "information." When it crosses a reflected beam, all the information in the interference pattern can be referred to the reflected beam alone.

Suppose, then, that we set up a situation as follows. A laser beam is made to shine obliquely on a piece of glass that is so treated as to allow half the beam to pass through, while the other half is reflected. The half of the beam that passes through continues to travel in a straight line until it passes through a rectangular opening. The half of the beam that is reflected strikes some object, and some of it is reflected again in such a way as to pass through the same rectangular opening.

The original laser beam (without a pattern) crosses over the richly patterned reflected beam and produces an interference pattern. All the information of the interference pattern would refer to the pattern of the reflected beam. If you were to look through the opening from the other side, you would see the object clearly despite the interference pattern formed with the laser beam.

Instead of allowing the eye to look through the opening, suppose we put a photographic film in the opening instead. In that case, a photograph will be taken of the interference pattern. All the light and dark areas will be recorded. The pattern is so fine, however, the alternate patches of light and dark so tiny, that nothing would be visible to the eye. The film would merely take on a slight grayness.

The successfully exposed and fixed film, carrying the interference pattern, is a "hologram."

Now suppose that a laser beam is made to shine down on the hologram from the same angle that the original laser beam shone down upon it when the hologram was formed. The laser beam passes through, illuminating the same interference pattern that had been set up when, originally, it had crossed the reflected beam. (Techniques had been developed whereby ordinary white light can be substituted for laser light at *this* stage.)

The same information exists now as before. That interference pattern contains within itself all the information of the pattern of reflection of the object originally illuminated by the other half of the beam.

If you look through the hologram illuminated by the laser beam, matters will be exactly the same (from the visual angle) as when you looked through the opening previously. You will see the object as perfectly and in the same way as though it were actually there. You will see it in its actual size, its actual appearance, and even in its actual three-dimensional characteristics. The hologram carries the whole message.

If, through the hologram, you are looking at the image of two chessmen, one partly behind the other, and you move your head in one direction, you will see more of the rear chessman. If you bend your head in the other direction, you will see less of it. Furthermore, you can focus on the nearer chessman, allowing the rear one to go somewhat out of focus; or on the rear one, allowing the nearer to go out of focus.

Of course, you can't do more to the image than you could to the real object; it would be unreasonable to expect to do so. When you look at the real object through a rectangular opening, there is a limit to how far you can see around an obstacle. If you move your head too far in one direction or another, you can move yourself out of the range of vision through the opening. The hologram fixes the opening, and you can't move beyond it. In the case of the real image, you can move around and behind it to look at its rear, but only at the expense of getting entirely away from the opening. Therefore you can't do that in the case of the holographic image.

Then, too, you can expect no surprises in a single *photograph* of a holographic image. A photograph taken of such an image is only a photograph, quite two-dimensional, and in itself has no holographic properties. However, you can take *different* photographs of the same holographic image; you can take photographs at different focuses and from different angles. The individual photographs may be ordinary, but a number of them taken together will give you a hint of the versatility of the holographic image.

There are some important ways in which a hologram, which has recorded an interference pattern, differs from an ordinary photograph, which has recorded a flat intersection of a reflection pattern. For one thing, there is no such thing as a hologram negative or a hologram positive. If all the white and dark areas of an interference pattern were reversed, it would still be the same interference pattern carrying the same information.

Then, too, a hologram is recorded without a lens. Different parts of the interference pattern are not focused on different parts of the hologram. Instead, every portion of the hologram is bathed in the crossing over of the two beams of light so that the interference pattern is recorded over and over again in every part of the hologram.

If you cut a hologram in half, you are *not* left with two halves of a complete picture. Each half of the hologram can be used to produce the complete holographic image. If you tore the hologram into ten ragged pieces, each piece could be used to produce the complete holographic image. If you scratched the hologram, the portion actually scratched would be spoiled, but all the rest would still produce the complete holographic image, with no signs of a scratch upon it. If you punched a hole in the hologram, you would still get an image with no sign of a hole. Dust upon the hologram would not interfere either, because the portions between the dust particles would still do the work.

There is, however, a limitation.

The interference patterns on the different parts of the hologram reinforce each other. The more repetitions there are of the patterns, the sharper and clearer the image is. As the hologram is torn into smaller and smaller pieces, or as it is subjected to more and more holes and scratches or to a thicker and thicker dust cover, the dimmer and fuzzier the image gets.

You can see this if you imagine someone writing his name with a very light pressure on a hard pencil and with a very shaky stroke. The name may be too dim and shaky to make out. If he repeats the process, though, writing his name over and over again in the same place, there will be repeated places where the pencil strokes will cross each other and where the result will be a darkening.

In the end, after hundreds of repetitions, there may be a gray area around the main thrust of his pencil strokes, but the crossings will concentrate along the lines and curves the writer is trying to make. The result will be that his name will appear sharper, darker, more even, than any single pencil stroke would have made it.

If you then imagine the pencil strokes removed one by one, the whole name will still be there, but it will gradually grow dimmer and fuzzier, and that is analogous to removing more and more of the pattern repetitions on the hologram by tearing, piercing, scratching, or dusting.

What are the applications of holography? To what use can it be put?

It is not revolutionizing the world all at once, for there is more to technological innovation than the mere working out of a new concept. It has to be made competitive; the concept has to be translated into hardware that will do something not only better than before, but more conveniently, more simply, more cheaply, or all three.

For instance, holograms can be made of some structure in double exposure. The object being holographed is left unstressed the first time, and is placed under some stress the second. The difference in interference patterns shows up as unevennesses in the double exposure, and these can be interpreted as representing defects of one sort or another in specific places in the structure being holographed.

In this way, holography can be used to test airplane wings, for instance, nondestructively. However, such objects can also be tested by X rays and ultrasonic sound, and holographic techniques are not sufficiently better or cheaper or more convenient to cause a wholesale shift in the direction of holography as yet.

Holographic techniques could be used to produce three-dimensional effects in television or movies, but holography produces too much information for television or movie techniques to handle just yet. Holography must wait for the older systems to catch up to its excellences.

Nevertheless, holography can do some things right now that cannot be done otherwise. One interesting application is its use in deblurring fuzzy photographs. It was the desire to do this that led Gabor to the original invention of the technique with respect to electron microphotographs.

Photographs can be blurred through known failings in the system used. A laser beam can be passed through the blurred photography and an interference pattern can be produced that will cancel out the effects of the failings. A new photograph appears in which the blurring has been greatly reduced.

The technique has been applied very successfully to the photographs taken by electron microscope. Such deblurring extends the range through which electron microscopes can produce successful magnifications. By use of the technique, the double helix of virus nucleic acid was shown for the first time and, eventually, single atoms may be made out.

It is quite possible that holograms will be made use of for storing information, a use for which their three-dimensional properties are not needed.

For instance, holograms might replace ordinary photographic techniques in many cases, since the holograms would be insensitive to scratching and minor damage that would be sufficient to spoil ordinary photographic film. TV cassettes may therefore become holographic eventually.

Then, too, hundreds of pictures can be recorded on a single piece of film holographically, when the blank laser light is made to shine through the film at a series of angles, each differing slightly from the next. A whole series of different interference patterns are formed with a whole series of different objects. These can then all be projected as a laser beam shines on the completed hologram first at one angle and then at another. Image after image appears, and the Encyclopaedia Britannica might be stored on a hologram the size of a sheet of typing paper—a sheet that under ordinary photographic techniques could record but one image and no more.

This same ability to store enormous quantities of information may result in the development of holographic memories for computers.

It is, however, perhaps useless to attempt prediction too closely and to see what holographic techniques might do, for instance, to aid medical diagnoses or surgical methods. Technology often takes surprising turns. Holography is a versatile way of handling large quantities of information, and exactly how it may be applied could depend on ingenious discoveries that would prove as sudden, as unexpected, and as productive as Gabor's original inspiration in 1947 proved to be.

B · Physical Sciences

14 · The Hydrosphere

Anyone approaching our planet from outer space and penetrating the cloud cover is almost certain to get one overwhelming impression as he circles Earth: water!

The surface of our planet is covered by a shell of water called the "hydrosphere." The major portion of the hydrosphere is an ocean; *an* ocean. We may talk of the seven seas and give separate names to what we call the Pacific, Atlantic, and Indian oceans. This, however, is just geographic convention, which we stick to for convenience and out of historical tradition.

The ocean, whatever we call its parts, is, in actual fact, a single, interconnected body of fluid, which contains 98 per cent of all the water on Earth.

It is a *large* body. The single ocean that bathes the shores of all the continents of the world covers 140 million square miles, or 70 per cent of the entire surface area of Earth. The ocean is three times as large in surface area as all the dry land of the planet put together; it is 40 times as large as the 50 United States.

And, of course, what we see is only the top of it. On the average, the ocean is 2.3 miles deep, and there are places where it is over 7 miles deep. The total volume of the ocean is about 300 million cubic miles. That means that if you built a square tank, 36 miles on each side, and poured all the ocean water into it, you would have to build the walls as high as the Moon in order to hold it all.

Ocean water is not pure water. You wouldn't expect it to be. Water is an excellent solvent, and can dissolve many different substances, which usually distribute themselves through the fluid as separate "ions." (An ion is an electrically charged fragment of an atom or molecule.)

Most of the substances that make up the rocky crust of the Earth are insoluble and do not dissolve. For that reason, the rocks, soil, and sand of dry land and of the ocean bottom remain essentially unaffected by the water that soaks them either permanently or only now and then.

Nevertheless, there are minor constituents of Earth's solid crust that are soluble and will dissolve in the ocean. The most common of these soluble constituents is sodium chloride, better known to most of us as table salt. Consequently, the ocean is a huge salt solution and is, in fact, 3.45 per cent salt. This means that there is some 54,000 trillion tons of salt dissolved in the ocean, and if all of this could somehow be removed and spread out evenly over the 50 states, it would make a heap 1½ miles high.

The solids contained in the ocean are not exclusively sodium chloride. About one seventh of the solids include ions containing every element on Earth—some present in greater quantity, some in lesser.

The ocean contains as part of its normal content even such substances as uranium and gold. In every ton of ocean water there is about one ten-thousandth ounce of uranium and about one five-millionth ounce of gold. This is so little uranium or gold in so large a quantity of water that it isn't practical to try to concentrate the metals and extract them from the water. Still, the ocean is so huge that the total amount present in solution at even these tiny concentrations is large. The ocean contains a total of 5 billion tons of uranium and 8 million tons of gold.

The ocean is unevenly heated. The surface of the tropic ocean is warmed, expands, and spreads outward north and south. Cooler water from north and south moves toward the tropics at deeper levels to balance this. The uneven heating of the ocean, the changing density of water with temperature, and the rotation of the Earth all combine to produce large currents in the ocean, swirling in enormous circles in some places, sinking in others, rising in still others.

The net effect is to keep the ocean well stirred up and the substances dissolved in it evenly spread out.

This is particularly important with respect to the ocean's content of the gases found in air. The oxygen and nitrogen of the air will dissolve in seawater to a slight extent. If the waters of the ocean were absolutely motionless, this solution would take place only at the surface. The lower reaches of the sea would get only the gas molecules that happened to blunder lower in their random motions. This would not be much.

As it is, though, the ocean currents carry dissolved gas through the sea to all parts and to all depths. The total amount of oxygen dissolved in the ocean is about 17 trillion tons, or about one three-thousandth of the total oxygen present in Earth's atmosphere.

This is enough. Ocean life depends on oxygen (as we do) and gets its oxygen from the quantity dissolved in the ocean. This means that the entire ocean and not just the surface layer is hospitable to life.

Still, the oxygen content of the ocean is not spread out entirely evenly. The amount of gas that can be dissolved in water varies quite a bit with temperature. The lower the temperature, the more gas will dissolve, so that the frigid waters of the polar areas can dissolve twice as much oxygen as can the warm waters of the tropics. It is for this reason that it is the cold waters of the Earth that represent the great fishing grounds and why huge whales, which must find food in ton lots, haunt the polar regions.

Life forms require not only oxygen but also a number of other substances dissolved in the ocean. The ions containing nitrogen and phosphorus (nitrate and phosphate ions) are an example. Living organisms incorporate such ions into their own tissue structure, leaving very little to remain free in the water.

The sea organisms eat and are eaten, but there are always some organisms that die without being entirely eaten. As a result there is a constant drizzle of dead tissue sinking slowly downward, and it is this drizzle that supports life in the ocean depths.

The downward-drifting bits of dead matter carry with them the nitrates, phosphates, and other ions making up the tissue, removing them from the surface where they are needed. Such ions are restored to the surface, however, by the upwelling of water in various places.

The Sun's warmth also has a profound effect on the ocean in ways other than that in which it produces water currents. About 380 trillion tons of seawater are evaporated by the Sun's warmth every year. If this evaporation went on indefinitely, the ocean would be entirely dry in about four thousand years—but, of course, the water is replaced.

The amount of water vapor that air can hold varies with temperature. At 86° F the air will hold seven times the quantity of water vapor that it will at 32° F. As a result, if the humid, vapor-bearing air over the warm tropic seas cools off, either because it moves north or south, or because it moves up into the cooler heights of the atmosphere (or both), the water vapor cannot remain in the air. It condenses into clouds of droplets, and eventually falls as rain.

The rain balances the evaporation, and the total size of the ocean does not vary appreciably from year to year. There is always some water vapor in the air, however, and the total amount present on Earth at any given time is about 14 trillion tons, or about one hundred-thousandth the quantity of water in the ocean.

It is only the water of the ocean that is evaporated, of course. The solid content remains behind. There are some tropic sections of the sea that are in areas where little rain falls and that open to the remainder of the ocean by constricted passages through which water does not pass freely enough. Such sections, notably the Red Sea and the Persian Gulf, tend to have less water and more dissolved material than the rest of the ocean. They are up to 4.0 per cent salt.

Not all the rain that falls, falls back upon the ocean. About 130 trillion tons of water fall each year on the land surfaces of the Earth. This is "fresh water," for none of the salt content of the ocean has come up with the vapor that has fed the rainfall. Rainwater is, in fact, almost completely pure water. It absorbs some of the gases of the air and may pick up a little dust, but that is all.

Some of the rain that falls on land evaporates again (and falls as rain again eventually). Much of it percolates down into the soil until it reaches some impermeable rocky layer above which it accumulates. If a well is dug deep enough, it will penetrate below the water level in the soil and will therefore contain water.

The water level reaches the surface in places where that surface is low enough. There it bubbles out of the soil and moves down-hill. The result is a river, which usually reaches the ocean at last. On its way it may pass through large-area depressions that will fill up with the river water and form ponds or lakes.

It is upon this fresh water in rain, in the soil, in rivers, and in lakes that most forms of land life (including, most particularly, man) depend for drinking water. The total amount of fresh water in liquid form on Earth is 560 trillion tons. This is about four years' worth of rainfall and is about one twenty-eight-hundredth the weight of the ocean. About one tenth of the Earth's supply of fresh water is found in the Great Lakes of the North American continent.

Some of the water vapor in the atmosphere falls as snow if the temperature is low enough. During winter in the temperate zones there is considerable snowfall, but the snow melts when warmer weather comes. In the polar land regions and at high elevations anywhere, the snow does not melt but accumulates.

The permanent continental ice that exists on Earth amounts to some 25,000 trillion tons. Nine tenths of this covers Antarctica, and most of the rest covers Greenland. Only minor quantities are found elsewhere.

The ice doesn't accumulate forever, of course. Under its own weight, it begins to flow, and these ice rivers, or "glaciers," slowly move to the ocean, where pieces break off that then float in the water as icebergs and, reaching warmer waters, slowly melt; or else the glaciers reach warmer land areas and melt there.

If all the ice could be melted and kept on land, it would represent an accumulation of fresh water 44 times as great as that present in liquid form now. The melted ice would not, however, remain on land, but would flow into the sea and raise the sea level by about 200 feet, with results catastrophic for man.

The liquid fresh water that annually finds its way to the sea by way of flowing rivers amounts to some 30 trillion tons. (As much as one sixth of this entire quantity may be contributed by the Amazon River alone.) River water is no longer as pure as rainwater, of course. Percolating through the soil, river water has dissolved small quantities of various substances.

River water contains only about 0.01 per cent dissolved matter, but that still means that every year, the rivers carry 3 billion tons of dissolved matter into the ocean. If this had been going on in the past, it could mean that the ocean, beginning as fresh water, had become as salty as it is now in 20 million years.

We cannot reach that conclusion, however, because we don't know at what rate rivers brought dissolved material to the oceans in the past. Furthermore, there are ways in which the ocean can give up its solids. The wind carries ocean spray far inland. There the water evaporates, but the solids stay behind. Then, too, shallow arms of the ocean evaporate completely, leaving their salt content behind as the vast salt mines that exist here and there on land. It may be then, that the salt content of the ocean stays fairly steady, year after year, despite the rivers' contribution.

Not all the river water, by the way, reaches the ocean. Some of the water finds its way into a hollow depression surrounded by higher ground on all sides. The water accumulates into an inland sea and does not leave, except by evaporation. The largest such inland sea on Earth is the Caspian Sea.

Since the water of an inland sea leaves only by evaporation, the solids are left behind. The river water pouring into the inland sea brings dissolved material and leaves it there. Since the inland sea is much, much smaller than the ocean, the dissolved material quickly accumulates to a high level. Such bodies of water as the Great Salt Lake in Utah and the Dead Sea in Israel are up to 25 per cent dissolved salt.

Earth's hydrosphere, then, is an extremely complex system, made up of the ocean in balance with water vapor, fresh water, and ice. The delicate balance, involving salt content and composition, oxygen supply, ocean currents, evaporation, rainfall and snowfall, ice accumulation, fresh water drainage, and the myriad activities of living organisms, is something we do not understand in all its details as yet.

But we do know that life depends on that balance and that small shifts one way or the other may wreak tremendous havoc on various life forms, including ourselves. We know also that man has interfered with this balance ignorantly and thoughtlessly, and that in the twentieth century, his ever more high-powered technology has made it possible for him to interfere with this balance more intensely.

It is man's effect on this balance that is now beginning to concern him deeply. What *is* the effect, and if it is harmful, how may it be reversed? That is a question that is a life-and-death one to all of us.

15 · Fresh Water

Suppose you drink two quarts of water a day, as doctors say you ought to. In the space of a year, you will have downed 730 quarts of water or, since there are 1,056 quarts to a cubic meter, you will have consumed about three-fifths of a cubic meter of water in the time from January 1 to December 31.

Of course, you use water for other reasons about the home, for anything from washing the dishes to watering the lawn to bathing. All told, the average American consumes, at home, 200 cubic meters of water per year.

Then, too, the water we drink is only the excess we require over the water content of food. We drink milk and fruit juices, and even solid food has a high water content. Water is therefore required in agriculture to keep plants and animals alive, and in considerable quantities, too. To grow a ton of wheat requires 8,000 tons of water spread over the growing season. What's more, industry is a lavish user of water. To make a ton of steel, for instance, requires 200 tons of water.

All told, the water needs of the United States comes to 2,700 cubic meters per year per person. Each of us uses 4,000 times as much water, directly or indirectly, as we drink.

In primitive regions, where industry is negligible and agricultural methods are simple, water needs can be satisfed by 900 cubic meters per year per person, and the average figure for the world as a whole might come to 1,500 cubic meters.

But the world is a watery planet, and if all the water on Earth were divided equally among the population (which is currently nearly 4 billion), then each of us would be the proud possessor of 375 million cubic meters. We would have far more than we could possibly know what to do with.

Yet wait. All the uses of water require it to be free of salt. Whether it is for drinking, for bathing, for cooking, for agriculture, or for industry, what we must have is fresh water. We cannot drink seawater, nor wash clothes with it, nor irrigate the soil with it, nor run our industrial plants with it. And, as it happens, a little over 97 per cent of all the water on Earth is inconveniently salty seawater.

What is left, the fresh water, if divided equally, would come to 10.25 million cubic meters per person. And not even all of this is available for use. Nearly three quarters of it is in the form of permanent icecaps, covering 10 per cent of the land surface of the world.

The liquid fresh water on Earth comes to about 2.4 million cubic meters per person. It is this that really makes life possible on land. If you were to prepare a map of the Earth in which various regions were darkened according to the quantity of liquid fresh water available, and another in which they were darkened according to population density, you would find that the two maps were the same.

Of course, fresh water, as it is used up, must constantly be replenished by the one great source of fresh water—rainfall—and that is unevenly distributed in time. There are rainy seasons, when the water supply builds up to more than is required and when much is unavoidably wasted; and dry seasons, when the water supply sinks dangerously toward a level of inadequacy. Then, of course, there are always droughts (or, for that matter, floods), which can be catastrophic and sometimes are.

As the population has risen to its present all-time-record height, and industrialization has expanded to its present all-time-record state, the need of fresh water per capita has increased, and the rise in the total need of fresh water has even further increased. Each year, therefore, it has become easier for drought conditions to become catastrophic.

We can approach the situation from another direction. Assuming an average need of 1,500 cubic meters of fresh water per year per person, the total amount of fresh water required by mankind in the space of one year is 5,550 billion cubic meters.

Earth's population is increasing by about 75 million people a year just now (and this rate is slowly rising). If we assume a continuing per capita need of 1,500 cubic meters per year and ignore the certainty of still further industrialization increasing that figure, then mankind's requirement for fresh water is increasing at a little over 100 billion cubic meters each year. If present trends continue, the total demand is sure to be twice what it is now by the year 2000 at the latest.

The supply of fresh water is refreshed by a total annual rainfall on this planet of 350 trillion cubic meters. This rainfall is well over 60 times the quantity used by mankind. That may seem to give us leeway, but not as much as we might think.

In the first place, much of the rain falls on the ocean, and that is simply a short circuit. Fresh water is formed, then restored to the sea before it can be used. The sea does not even need that rainfall, for if all the rain fell on land surfaces, the rain would still be quickly returned to the sea. (Of course, if *all* the rain fell on land, there would be flooding problems here and there.)

Then, too, some of the precipitation falls as snow on the permanently icy areas of earth. Such precipitation is restored to the sea in the form of icebergs, and the fresh water represented by those icebergs goes unused.

What is left of the precipitation falls as rain (or as snow that eventually melts) on those portions of the land surface that are not under ice. Some of this evaporates before anything else happens to it (and this is most likely to take place in hot, arid regions, which need liquid water most). The evaporated water must then run the gantlet a second time and may fall, that second time, on the sea or on the icesheets.

The water that is not evaporated sinks into the soil and becomes "ground water," leaking down through many feet of soil and of permeable rock until, inevitably, at some depth, impermeable rock is reached. The ground water, resting on this impermeable rock, reaches some height within the ground that is held in balance by the addition of further rainfall and the subtraction of downhill runoff that eventually ends in the sea.

The height reached by the ground water is the "water table." If one sinks a hole deep enough almost anywhere, one penetrates below the water table, and the hole fills with water to that level. From such a "well" one can apparently draw water indefinitely, for the level is likely to remain steady. (Under conditions of severe drought or impassioned overuse, the water level will, however, inevitably drop.)

The layer of impermeable rock on which the ground water rests is itself not level. It rises and dips, more or less with the configuration of the surface. The surface of the ground in a valley may dip below the water table in neighboring hills, in which case the water can rise to ground level in that valley. It won't, however, if there is a surface layer of impermeable rock over the water table. If, in that case, a well is drilled through that upper layer, water rises with force and comes shooting out of the surface. These are "artesian wells," so called from wells of this sort dug in the French province of Artois.

The total supply of ground water is great, coming to about 2.3 million cubic meters per person, or 1,500 times what each of us requires in a year. What's more, it is everywhere, since even in regions receiving little rainfall, ground water seeps in from adjoining areas that receive more. There are great amounts of groundwater, for instance, deep under the surface of the Sahara Desert. The problem is, of course, to reach that groundwater.

The groundwater may come up of itself. Where the ground dips so that its surface is below the water table and there are no impermeable rocks to get in the way, water bubbles out of the ground in the form of a "spring." The water trickles downhill where, if the surface grooves so that it continues to be below the water table, other ground water is added so that the spring swells to a stream and then to a river. Where there are extensive depressions, the water accumulates into ponds and lakes.

(The population of the eastern United States is accustomed to steadily flowing rivers fed by a stable water table. In other areas, where rainfall is very unevenly distributed in time, the water table sinks markedly in the dry season and rises in the wet season, so that rivers can vary between a trickle and a flood.)

The free water of rivers and lakes is the most convenient form in which fresh water is found. As sources of drinking water, as support for ships, as a source of energy, they are worth their weight in—water. The total amount of such free water on Earth is 54,000 cubic meters per person—35 times the individual requirement per year and even 20 times the particularly high annual requirement of Americans.

But, of course, not all the river water and lake water is used. Some of it is constantly evaporating, and much of it pours off into the sea with very little use having been made of it. The Amazon River, which drains the wet, equatorial areas of South America, is the largest river in the world and discharges into the sea in a year enough fresh water to supply 1,800 cubic meters for every person on Earth. That is enough, in itself, to supply all the needs for everyone, if it could be saved and distributed. In actual fact, virtually none of the Amazonian water supply is used by man.

Considering, then, the actual availability of fresh water, and the amount that, for one reason or another, doesn't get used, we end with the realization that despite the watery nature of our planet, we are pushing the limit. We can go very little farther if we are merely content to continue taking what fresh water comes our way without trouble. We must begin to plan.

1. Since the ultimate source is the ocean and since all the water evaporated from it returns to it, we will do no harm if we try to speed the evaporation. By desalting the ocean one way or the other, we can obtain a direct and essentially unlimited supply of fresh water. To do so costs energy, however, and methods must be devised to make enough energy available (solar energy perhaps or tidal energy) to do the job in sufficient quantity to be useful.

2. Since rain falling on the ocean is completely wasted, anything that will encourage rain on land (with limits short of disastrous flooding) is all to the good. Land can be built up at the expense of the ocean in certain favorable areas, as in the Netherlands, and can receive rain that would otherwise fall on water. Cloud seeding, or still more sophisticated methods to be devel-

oped in the future, may succeed in directing rainfall precisely to those areas where it is most needed.

3. Since snow falling on the icecaps is of no use to us, we might develop methods for exploiting the frozen fresh water of icebergs instead of letting them melt uselessly into the sea.

4. Since evaporation from lakes and from the soil is an important route for fresh-water loss, methods for reducing evaporation might be developed. Single-molecule films of certain solid alcohols might be layered over exposed fresh water to keep down evaporation.

5. Since the river runoff into the sea represents a waste of fresh water, every effort should be made to make use of the water before allowing it to reach the sea.

6. Since the groundwater supply is, in total, some forty times as great in volume as the water found in rivers and lakes, and since the groundwater supply is much more widely spread, every attempt should be made to exploit that groundwater efficiently.

7. Since the supply of fresh water is increasingly critical, every effort should be made not to render any of what exists useless. This means that the pollution of lakes, rivers, and groundwater must, at all costs, be held to a practical minimum.

8. Finally, since the population increase raises the requirement for fresh water each year, as does the steady advance of industrialization, we must understand the limits of growth. If population is not controlled, the continuing growth of the mass of humanity will eventually overtake any increase we can bring about in the fresh water supply; and the collapse that follows will be the more catastrophic the higher the population level from which it begins. If the advance of industrialization is not to help bring about this catastrophe, it must be guided at every step by a prudent consideration of how best to conserve such fundamental resources as fresh water, and a foresighted consideration of consequences of any action to society and, even more fundamentally, to the ecology.

Given ingenuity, good sense, and (most important of all, perhaps) good will, we can yet create a flourishing and happy planet, but the time of grace during which we can accomplish this is growing short.

16 · It's About Time

Who cares what time it is? Who cares how old one is? Exactly, anyway?

Until quite recently in human history, nobody did. When a girl began to menstruate and show definite breasts, she was old enough to marry. (Juliet was not quite fourteen years old when she caught Romeo's eye.) When a woman reached the menopause, assuming she had survived her numerous childbirths, she was old.

Beyond that, what more did one have to know about a woman's age?

As for the male sex, when a body grew a beard, he was a man. (Before that he was merely a beardless youth—a caustic gibe in early times.) When the hair and beard began to streak with gray, he was old. The gray often appeared first at the place where hair and beard met, and that place is called the "temple," from the Latin word for time.

And what else is necessary?

Yet nowadays we know the exact day of our birth, and every year we watch the birthdays pass, and all the anniversaries, and the hours and minutes and even the seconds.

How did all this modern concern with time come about?

As usual, the very beginnings vanish into the mists. No one knows how the notion of time began. No one can ever know.

We can only suppose that somehow the concept arose that first it was day, then it was night; first one ate and then one was hungry again. The concept of "first—then" implies the passage of time.

But mere passage must be followed by the problem of duration. If it is now night, how long will it be till day? If you have now eaten, how long will it be till you are hungry? There was no way to answer such questions at first except by the obscure sensation of duration within you. You seem to wait a long time or a short time, and you learn by experience how long you may expect to feel you will have to wait.

The trouble is that the sense of duration is not something different people can agree on. It is not even something the same person can experience in the same way at different times. If one person is waiting through a cold, wet night in the open, without a fire, the dawn seems very long in coming. To another person who is in a dry cave, stretched out near the glowing embers of a fire, the dawn comes quickly—all too soon, perhaps.

Which was it *really*, then? Was the night unusually long or unusually short?

The obvious answer might be: "I don't care what *you* say; it was a long, cold, miserable night as far as *I* was concerned." And, in a way, that's right; it's your own life you have to live.

If human beings lived solitary lives, each on his own desert island, it might be natural for each to live according to his own notion of time. The trouble is that people must co-operate. It is far more convenient for a whole family to eat at the same time, for whole tribes to adjust working, eating, and sleeping schedules. Individual sensations of duration must give way to the general consensus.

And how about traveling? Early man was a hunting creature and had to roam widely in search of food. How distinguish the distances between various points in the range, when the very concept of distance was vague? You might point to a distant object—a mountain peak, perhaps—and say, "Head toward that, and keep walking for a certain duration of time."

Ah, but how do you express the length of duration? The solution must occur to any man, however primitive. It seems unavoidable that he would say, "Walk three days in that direction and on the fourth you will reach the place I am telling you about."

Some days or nights might seem unusually long or unusually short, for some personal reason. Then, too, days were distinctly shorter for everyone at some times of the year and longer at other times; so were nights. Nevertheless, men must have decided that on the whole, a day plus the following night represented a duration of time that was always the same. At least, the assumption that the duration was always the same seemed to work.

There were important rhythms in nature that were much slower than that of day and night. There were periods of rain or of rising rivers, when food was plentiful; and there were periods of drought, when the food supply dwindled. Or in some parts of man's range, there were periods of warmth and plenty, when days were long and nights short; then periods of cold and snow, when days were short and nights were long.

It was one of the great discoveries of mankind that, in general, such periods did not come haphazardly, but followed each other in roughly regular alternation. Mankind discovered the seasons!

This discovery may well have been one of the most important in the history of human philosophy. It gave rise, perhaps, to the thought of the inevitable. Summer *had* to be followed by winter; feast by famine. Nothing man could do seemed to alter those facts. Perhaps it was through musing on those facts that somebody first developed the notion that man *had* to die. Even with all possible good fortune, with escape from enemies, from wild beasts, from famine and disease, each person nevertheless *had* to age and die.

But then, as surely as summer declined into the fall and as all nature died with the winter, so that winter was followed by spring and a rebirth. Would that not give men the hope that even though death were unavoidable, there would yet be a rebirth somehow, and out of that might there not have come the first glimmerings of religion?

Again, there arises the question of duration. It was not enough to know that spring was coming and that if one merely waited, the snows would disappear. *When* would this happen? How long, O Lord, how long?

To count days was tedious, and there were hundreds of them from one spring to the next. But what about the Moon and its constantly changing shape?

The Moon would appear first as a thin crescent in the western sky just after sunset and then, from night to night, it would grow thicker and shine higher in the sky, till finally it was a complete circle of light, a full Moon. Then it would begin to narrow again, appearing later and later in the sky until it would be thinned down to a crescent rising just before dawn.

A couple of days would pass and a *new* Moon would make its appearance as a thin crescent in the western sky just after sunset, and the whole process would be repeated.

Then one day, some prehistoric astronomical genius, counting the days from new Moon to new Moon, found a constant interval. There were, on the average, 29 and one half days from new Moon to new Moon, and there were about 12 new Moons from spring to spring.

It was easy to count 12, and it was a great and useful discovery to know at any time when the next spring would come. There were 12 new Moons in a cycle of seasons or, as we would say, 12 months in a year.

Twelve was such a convenient number, too, because it could be divided evenly into halves, thirds, fourths, and sixths. No other number so small could be divided evenly in so many ways, and to primitive men with just a beginner's grasp of arithmetic, that was important.

The fact that 12 was so useful in keeping time and so convenient in doing arithmetic gave it a mystic importance, so that we end up with 12 signs of the zodiac, 12 hours in a day, 12 inches in a foot, 12 tribes of Israel, 12 apostles, twelve people on a jury, and 12 of anything in a dozen.

Then women invented agriculture. (The men were too busy hunting and sleeping to pay attention to the weeds growing about the campfire, whereas women were for long ages the gatherers and cookers of plant food, and some must have gotten the idea of cultivating them deliberately.)

In order to grow grain properly, one had to know exactly when to plant in order that harvest time might come before the snows. As population increased and became utterly dependent on that harvest, counting new Moons became a matter of life and death.

To insure the correct count, each new Moon became a religious festival. The very word "calendar" (from the Latin *calare*, meaning "to proclaim") reminds us that each month began with a new Moon, loudly and ritualistically proclaimed by the priests.

There are many references in the Bible to the religious festivals that marked the new Moon. In the Eighty-first Psalm, third verse, for instance, there is the exhortation, "Blow up the trumpet in the new Moon, in the time appointed, on our solemn feast day."

There was one trouble, though. The count of 12 months in the year wasn't exact.

Suppose a new Moon came at the ideal moment of seed time in the course of a certain spring. The twelfth new Moon after that came just a little before the ideal moment of seed time. Each twelfth new Moon afterward came a little farther too soon, until, if you kept on counting strictly by twelves, you would find yourself trying to plant your grain while the ground was covered with a foot of snow.

Every once in a while, the priests would have to slip in an extra month to keep the calendar even with the seasons. This must have been annoying, for it upset the established order of things and introduced complications. Then, too, 13 could not be divided evenly by *any* number, so that it was difficult to speak of half a year or a quarter of a year.

It is very likely that this is the origin of the common belief that 13 is an unlucky number.

It would be a little more convenient, of course, if someone would devise a regular system for adding that thirteenth month, so that it could be carried through regularly and mechanically without too much trouble. This was done in Babylonia in about 550 B.C.

According to the Babylonian system, the years were counted

in groups of 19. In each group of 19, the third, sixth, eighth, eleventh, fourteenth, seventeenth, and nineteenth years had 13 months; all the rest had 12 months. In this way a particular new Moon might be a couple of weeks ahead of the season or behind, but never more than that. And it could all be done mechanically.

The Greeks borrowed this system from the Babylonians, and so did the Jews during the period when they were held captive in Babylon. The Jewish calendar to this day has this regular system for putting a thirteenth month into the year at fixed intervals.

It was the Egyptians who worked out an entirely different system, thanks to an accident of nature. The Egyptians, in their rainless, winterless country, depended for life on the rising of the Nile each year. But that rising occurred with much greater regularity than did the coming of the first rains in other countries, or the coming of the first frosts.

It was not long before the Egyptian priests realized that the floods came at intervals of 365 days on the average, whereas 12 new Moons counted off only 354 days.

Very early in their history, perhaps when the pyramids were being built, the Egyptians decided to go with the floods and to disregard the Moon. They made their months each last 30 days for a total of 360 days, then added 5 extra days and began the count over. This meant that a new Moon might come on any day of the month, but the Egyptians didn't care.

The Egyptians noticed something else. In their sunny land, it was easy to drive a stick into the sands and then notice the way its shadow shifted as the sun moved across the sky. They couldn't help but notice that at some times of the year, the noonday Sun was particularly high in the sky and the stick's shadow was unusually short. At other times, the noonday Sun was lower in the sky and the shadow was longer.

By following the change in length of the noonday shadow, it turned out that the noonday Sun rose and fell in the sky in a slow rhythm that lasted 365 days. The Nile floods and the season followed the sun, and what the Egyptians had was a "solar calendar."

The Babylonians also discovered this yearly cycle of the Sun but, without the regular Nile flood to guide them, they did so only after the system of new Moons had grown too sacred to budge. They therefore kept the "lunar calendar."

The Romans used a lunar calendar but did not use a fixed 19-year system for introducing the thirteenth month. Their priesthood consisted of political appointees, and the Roman high priest generally introduced a month whenever he wanted to keep his own party in power an extra length of time. The result was that by 48 B.C., the Roman calendar was in hopeless confusion.

In that year, though, Julius Ceasar was in Egypt. He was just passing through, he thought, but he found Cleopatra there and decided to stay a while. In Egypt, he studied its customs and had its solar calendar explained to him.

He thought it much simpler than the lunar calendar and less likely to be subjected to political manipulation. When he got back to Rome, he supplied the Romans with a new calendar. The Roman people didn't like to have their old calendar interfered with, but Caesar had a way with him, and when he said solar, it was solar.

In fact, he improved on the Egyptian calendar. Actually, the cycle of seasons lasts 365¼ days. To take care of that quarter day, Caesar arranged to have every fourth year extended to 366 days. This was a big improvement.

He also arranged to have the 5 extra days of the Egyptian calendar distributed through the months so that some had 31 days and some had 30. This was a change for the worse.

Caesar made sure that the fifth month, his own birth month, had 31 days, and it was renamed July in his honor. His successor, Augustus, had the sixth month, his own birth month, renamed August in *his* honor, and saw to it that it had 31 days, too. February, however, was cut down to 28 days because the Romans considered it an unlucky month.

This calendar of Julius Caesar, with its leap years, is still the calendar which, with minor improvements, we use today. It is a gift from Cleopatra, without whom Caesar would not have remained long enough in Egypt to learn a little astronomy.

While the Romans switched to a solar calendar, the Jews and Greeks held on to their lunar one. So did the early Christians, most of whom were either Jews or Greeks.

The Jewish holy days come on fixed days of fixed months, according to the lunar calendar. But that means they hop all over the place from year to year according to the solar calendar we use.

The holy days of the early Christians do the same. The chief holy day of the early Christians was Easter, which follows the lunar calendar and can therefore be anywhere from the end of March to late April in our solar calendar. It is a "movable holiday."

When the Roman Empire became officially Christian, the Church adopted the Roman calendar, but it then had to figure out exactly when Easter ought to be celebrated from year to year by the older calendar. Complicated systems involving the 19-year cycle of the lunar calendar were worked out, and this was a lucky thing, for it helped keep learning alive during the Dark Ages. You see, no matter how severely secular learning might be denounced by over-eager theologians, there always had to be *some* astronomy taught so that the date of Easter could be calculated, and once you allow a little learning, people grow curious for more.

To be sure, not everyone agreed on the exact system of working out the date of Easter. In the sixth and seventh centuries, the so-called Celtic Church, established in Ireland, competed strongly with the "Roman Church," and just about the greatest difference between them was the system used to calculate the day of Easter. For a while it seemed that the Church might be split in two on this subject, but then the Roman system won out.

Christian holy days that came into use after the Church had switched to the solar calendar remained fixed and did *not* move about from year to year.

In Roman times, the day on which the noonday Sun reached its lowest point and began to move upward again (the "winter solstice") was celebrated with great abandon. It meant that the spring would surely come back again and the Sun would not, after all, sink lower and lower and die altogether. It would be born again.

The celebration of the birth of the Sun was called the Saturnalia (after Saturn, a god of agriculture, since the birth of the Sun was essential for agriculture). On the Saturnalia, there was gaiety, present-giving, and brotherhood for all. It was, in general, a cheerful season of good will.

Since it was an idolatrous celebration, the Church was against it, but they couldn't wipe it out. It was too gay and beloved a holiday to abandon, so the Church made the best of things by adopting it. There was no indication in the Bible when Jesus was born, and it might just as well be at the time of the Saturnalia as at any other. The name was changed to Christmas, therefore, and at every December 25 we still celebrate the birth of the Sun (or the Son, if you prefer).

But whether the year be lunar or solar, how do you keep track of them?

The easiest way is to identify each by some event: That was the year of the big wind. That was the year in which Mary had her second child. That was the year Nixon was elected President.

The ancients did that. They identified the year as the year in which so-and-so was consul, or as the sixth year of the reign of Ptolemy II. The Bible, when it identifies years at all, does so in this manner, so that virtually no event in it can be certainly dated.

Even so important a date as the birth of Jesus is described as having taken place "when Cyrenius was governor of Syria" (Luke 2:2), and nothing more. But Cyrenius was governor of Syria in more than one year.

The apparently simple notion of numbering the years from some agreed-upon beginning came surprisingly late in history. Eventually, the Greeks numbered years from the first celebration of the Olympian games; the Romans numbered theirs from the legendary date of the founding of Rome.

The Christians used the Roman system at first, then took to numbering the years from the birth of Jesus. We use the latter system now over most of the civilized world, but it did not come into general use in Western Europe until the time of Charlemagne, about 800.

Even so, the system is not quite accurate. An early Christian theologian calculated the birth of Jesus to have taken place just at the beginning of what we now call A.D. 1, but the Bible says he was born in the reign of Herod, and Herod died in 4 B.C. It follows that Christ could not possibly have been born later than four years "before Christ," a curious consequence of the sloppy way in which the Bible tells time.

But let's go in the other direction. Days and months and years reached their present sophistication in 1582, when Pope Gregory XIII authorized the final refinement of the calendar (leaving out three leap years every four centuries). What about small periods of time of less than a day, however?

What was needed was some physical phenomenon that took place in regular intervals of time that were less than a day, but there were none such known.

You could have the Sun dial mark off the motion of a shadow, but the Sun shifted position during the course of the year, which complicated things.

By night there were the stars that moved around the Pole Star. By observing the position of the Big Dipper against landmarks, the starry sky became a gigantic clock. Thus in Act II, Scene 1 of Shakespeare's *Henry IV*, Part I, someone says that the Big Dipper "is over the new chimney, and yet our horse not packed." But the stars shift position through the year, too.

Besides, there are often clouds, which hide Sun and stars alike.

How about artificial phenomena to mark the passage of time? How about the slow dribble of fine sand through a narrow orifice; or the slow burning of a candle; or the slow and steady drip of water from a reservoir; or the slow descent of a weight attached to gears?

All these systems have been used, and all could tell time to the nearest hour, but no better. Even in 1582, when the calendar got its final polish, there existed no device anywhere on Earth that could tell time to the nearest minute.

But did anyone need such close timing? If so, why?

The answer came with an Italian scholar named Galileo. He grew interested in the exact manner in which objects fell. In the 1590s, he let balls roll down slanting grooves. In order to analyze the motion, he had to know the exact time it took them to go different distances.

He was the first scholar to want to know *the exact time* it took something to happen, and that was the moment in which modern science began.

But Galileo had no good way of determining exact times. He did his best by counting his pulse and by letting water drop through a small hole and weighing the quantity that came out. This was not very good, but it was good enough to allow him to revolutionize the science of physics.

Yet it was ironical that he was reduced to counting his pulse, for he had made an important discovery earlier that was the answer to time-keeping.

In 1581, when he was only seventeen years old, he was attending services at the Cathedral of Pisa and found himself watching a swinging chandelier. Air currents shifted it now in wide arcs and now in small ones, but to Galileo, it seemed that the time it took to swing back and forth was always the same, whether the chandelier moved a little or a lot.

He rushed home, experimented with what we now call a pendulum, and found himself to be right. For the first time, mankind had discovered a phenomenon that moved in a reasonably steady cycle that lasted less than a day—less than a second, if properly constructed.

Yet Galileo never put the pendulum to use as a time-telling device.

The Dutch physicist Christiaan Huygens did. In 1656, he attached a pendulum to the gears of a clock in such a way that the steady beat of the pendulum (kept in motion by the energy of slowly falling weights) made the hands move very evenly around the dial. The "grandfather's clock" he invented was the first really accurate timepiece ever made.

But still, who cared—except for a few weird scientists?

It turned out that merchants and seamen had to care. After the time of Columbus, increasing numbers of ships began to scour the oceans, and sea trade became more important every decade. In order to be able to find their way across thousands of miles of open water and end up where they wanted to go, ships had to know their own latitude and longitude at all times. It was easy to determine latitude by observing the position of the Sun. In order to determine longitude, however, it was important to know the exact time. Nor would a grandfather's clock help, for the beating of the pendulum would be instantly upset by the swaying of the ship.

Accurate time-telling became a matter of international concern. The British Government, particularly, found it necessary to be involved. It offered a prize of twenty thousand pounds to anyone who could devise an accurate timepiece that would stay accurate on board ship.

A clever mechanic, John Harrison, did the job. Beginning in 1728, he constructed a series of fine clocks, each better and smaller than the one before. The fifth clock was so small it could be held in the hand.

The British Parliament, in an extraordinary display of meanness, refused to pay up. Harrison couldn't force the money out of them till 1765, and then only with the help of a sympathetic George III.

After Harrison's time, men of substance began to carry pocketwatches. And once you could tell the time accurately, you made appointments by the minute and kept them by the minute, since how else could you demonstrate that you owned an expensive and reliable watch? Mankind became timebound.

New technological developments took advantage of that, too. Trains left the stations on the minute (and any decent person with a decent watch wouldn't miss it) and arrived, sometimes, on the minute as well. Then came radio and television, with programs that started on the *second*.

Now we have watches based not on moving pendulums, springs, or tuning forks, but on the vibrations of atoms themselves. The atoms vibrate so steadily that we can measure the amounts by which the movements of the heavenly bodies are uneven. We can tell that the length of the day varies very slightly from time to time and is slowly getting longer over the aeons.

So timebound humans have a new way of life. They must get to work at a fixed time and leave at a fixed time, so that there are "rush hours." The whole city has its fixed lunch hour, and the restaurants are either crowded or empty. The department store opens at a certain time, and on the days of a sale, thousands know exactly when and are waiting feverishly for a door to swing back.

And a woman knows, to the minute, when she will turn thirty.

Maybe Galileo should have kept his mind on his prayers that day in the cathedral.

Chapter 16 • AFTERWORD

Because I am so obviously prolific, the rumor sometimes arises that I never receive rejection slips. (No, I don't start those rumors myself.)

Anyway, it's not true, and the foregoing essay is an example of one that was rejected. A woman's magazine (no names, please) commissioned it, and I warned them that it would contain neither sex nor topicality and they said they didn't care and I wrote it and they said they couldn't use it because it contained neither sex nor topicality. Oh, well, that's the writing game!

What I do in such cases is to place the essay to one side and wait.

Half a year later, I received a letter from a gentleman who was planning a new magazine called *The Construction Man.* He asked if I had something he could use, and I said that the only thing I had on hand that was even remotely suitable had been written for a woman's magazine. He said he wanted to see it anyway. I sent it to him, and he accepted it and placed it in the featured spot in the first issue of the magazine.

Anyone reading the essay now is liable to notice that I seem to be addressing women, and anyone reading the copyright acknowledgment will see it appeared in *The Construction Man.* If that puzzled you, you need now be puzzled no longer.

17 · Overflowing the Periodic Table

Between 1898 and 1913, about forty chemical elements were discovered, more than in any like period before or since. Of course, most of those elements turned out not to be elements. In fact, only five survived. Nevertheless, the names of all forty are still in the books to this very day to confuse and confound the student.

The trouble was that, prior to 1913, no chemist had ever heard of isotopes. In the innocent days preceding World War I, each element was considered to have its own particular type of atom, with a particular atomic weight. All atoms of a given element had the same weight, and this weight was different from that of all atoms of all other elements.

Each element had its own unique set of properties and, conversely, whenever you found two elementary substances with different sets of properties, you were obviously dealing with two different elements.

Obvious, yes—but wrong!

What is really unique about an element is its atomic number and not its atomic weight, but the atomic number wasn't discovered until 1913, either. The atomic number of an atom is equal to the number of protons in its nucleus, which is, in turn, equal to the number of electrons in the outer reaches of the atom.

It is the number of electrons in an atom that determines its chemical properties. It follows, then, that all atoms with the same atomic number have the same chemical properties, and vice versa.

The atomic weight of an atom (which was the property that chemists, prior to 1913, had thought to be crucial) is equal to the number of protons plus neutrons in the nucleus. (The neutrons don't affect the number of electrons in an atom, hence don't affect the chemical properties.)

Suppose one atom has x protons and y neutrons in the nucleus and another atom has x protons and z neutrons in the nucleus. Both have the same atomic number, x, so both have the same chemical properties. The atomic weights, however, are different. The first has an atomic weight of $x+y$; the second has one of $x+z$. The two atoms are isotopes of the same element.

As a matter of fact, most of the elements are made up of two or more isotopes. In each case, the chemical properties of the isotopes are just about indistinguishable, so no one noticed them.

But then, in 1896, radioactivity was discovered. Different isotopes of an element might and did turn out to have radically different radioactive properties.

The radioactive properties of an atom, you see, depend on the stability of the nucleus, and that depends on the number and interrelationships of both protons and neutrons in that nucleus. Add one neutron to a nucleus and though the chemical properties of the atom as a whole are unaffected, the balance of the nucleus may be radically altered. The nucleus may be changed from a stable one to an unstable one, or vice versa. It might alter the degree of instability.

Between 1896, then, when chemists first started studying and measuring the radioactive properties of elements, and 1913, when the British chemist Frederick Soddy had completely worked out the theory of isotopes (without the proton and neutron details, since neutrons weren't discovered till 1931), what happened to chemistry shouldn't happen to a dog.

To begin with, two elements were known that turned out to

be radioactive. In 1896 the French chemist Antoine-Henri Becquerel made the first discovery when he found that a uranium compound would fog a photographic plate in complete darkness. In 1898, Marie Sklodowska Curie (a Polish chemist working in France) discovered that thorium was radioactive. In 1896, uranium had been known for 107 years. In 1898, thorium had been known for 69 years.

Next, Madame Curie made an important observation. She was working with pitchblende, an ore that contained uranium atoms. She knew how much uranium was contained in a given amount of that sample of ore. She knew how much radioactivity was to be expected of that amount of uranium.

What she didn't know was why the pitchblende should be producing four to five times as much radioactivity as was to be accounted for by the uranium content. Conclusion: There was something in the pitchblende, besides uranium, that was radioactive.

But she knew practically everything that was in the pitchblende, and none of it was radioactive except for the uranium. Further conclusion: The extra radioactivity came from some element that was present only in traces, and if such a trace element produced all that radioactivity, it had to be a devil of a lot more radioactive than uranium.

Marie Curie and her husband, Pierre, started looking. In July 1898, they discovered a new element, polonium, and in December 1898, they discovered another new element, radium.

So far, so good, but chemists couldn't stop. For fifteen more years, they kept discovering elements.

In 1900, the British chemist Sir William Crookes, found that he could isolate from a solution of a uranium compound, a substance that was more active than uranium. Since the substance displayed radioactive properties that were different from those of any known element, it had to be a new element in the light of views then current. He hadn't the slightest idea of any of its other properties, since radioactivity could be studied quite adequately in amounts of substances too small to test directly by any ordinary chemical method. So he called it simply uranium X.

But then, other men isolated other substances from uranium, also with new and different radioactive properties and therefore considered to be new and different elements. What Crookes had called uranium X became uranium X_1, and another substance became uranium X_2. There was also a uranium Z. Ordinary uranium was called uranium I, and there was also a uranium II.

Each of the uraniums was given a different symbol, as was their right. (After all, weren't they different elements?) Uranium I was UI, and uranium II was UII. There was also UX_1 and UX_2. A uranium Y (UY) was discovered, and furthermore there was a variety of uranium that gave rise to a different set of radioactive substances altogether. It was called actino-uranium, and its symbol was AcU.

These names sound provisional, you see, and the symbols do, too, because they're not like the symbols for other elements. The hope was that when enough was collected of each of these new radioactive elements to enable other, more usual properties to be studied, why, then, other and more reasonable names would be given them.

Occasionally, some chemists jumped the gun. For instance, Kasimir Fajans (an American chemist) and O. H. Göhring (a German), who discovered uranium X_2 in 1913, called it brevium, from the Latin word meaning "short," because it existed for such a short time. (It had a half-life of just over one minute.) The name didn't stick, however.

Another example of that involves the American chemist B. B. Boltwood and coworkers who, in 1904, first showed that radium was produced as one of the products of uranium breakdown. In 1907, Boltwood further showed that there was an element in between. It was formed from uranium and broke down into radium and had properties different from either. He called the new element ionium (symbol, Io) because, like all radioactive elements, it produced ions in the substances it irradiated. That name stuck for a while.

The German physicist Friedrich Ernst Dorn had shown in 1900 that when radium broke down, a gas was produced. Since the gas emanated (that is, flowed forth) from radium, it was first called radium emanation. However, since the gas was shown to

resemble the inert gases (argon, neon, etc.) in chemical properties, it was given a name of similar form, niton (from a Greek
word meaning "shining"). Later, radon was substituted, in order
to show the relationship to radium, and that name actually became official.

Thorium also gave rise to a gas with inert gas properties. This
was first called thorium emanation and then thoron.

Actino-uranium was found to break down to protactinium and
then to actinium. In fact, actinium was discovered first. The
French chemist André Debierne discovered it in 1899 and named
it from a Greek word meaning "ray." In 1917, Otto Hahn (German) and Lise Meitner (Austrian) discovered protactinium and
named it that because it was the ancestor of actinium. Later still,
a variety of uranium was found to produce protactinium, and that
was why the name actino-uranium arose.

However, the point is that actinium in breaking down gives
rise to a gas that was called actinium emanation or actinon.

That's three radioactive gases—radon, thoron, and actinon—
each with different radioactive properties, and hence, to all appearances, three separate elements. What's more, all three were
inert gases, and there just wasn't any room in the periodic table
for three new inert gases. Chemists were turning good and sick by
then.

Between 1902 and 1904, the British physicist Sir Ernest Rutherford found that radon broke down in a series of steps that
could be followed by changes in characteristic radioactive properties. He called the individual products radium A, radium B,
radium C, radium D, radium E, radium F, and radium G. Radium F, it turned out, was the polonium that had been discovered six years earlier by the Curies. Radium G was, finally, a
stable compound. In fact, it was lead.

Each of these different substances possessed its own symbol—
RaA, RaB, RaC, and so on. When further studies by the team of
Hahn and Meitner discovered two radioactive substances between radium C and radium D (this was in 1909), what were
they to be called but radium C' and radium C".

The whole succession of elements worked out between 1900
and 1909, from uranium down through the other uraniums, io-

nium, radium, radon, and then the remaining radiums, was called the uranium-radium series.

As a result of the work of a number of chemists and physicists of all nations (I have been stressing nationalities in this chapter to show how international scientific research is), it was shown that actinon and thoron both broke down in a series of steps similar to that of radon. As a result, the substances actinium A, actinium B, actinium C, actinium C′, actinium C″, actinium D, thorium A, thorium B, thorium C, thorium C′, thorium C″, and thorium D were discovered. One set belonged to what was then termed the actinium series, the other to the thorium series. Three independent radioactive breakdown series were thus worked out.

Furthermore, Hahn started the research that ended in discovering the substances produced in the thorium series between thorium itself and thoron. These included mesothorium I, mesothorium II, radiothorium, and thorium X.

By 1913, the horrible facts of the matter were this. There were about forty types of radioactive substances known, each one different from the rest. By all the rules of chemistry they seemed to be forty different elements. All of them had to fit into the periodic table between element No. 82 (lead) and element No. 92 (uranium).

Slice it however you wished, that meant trying to squeeze forty elements into a place where only nine elements could go. Something had to give—either the periodic table or chemists' notions about elements. Down went the notions. Up went new and better notions.

The first hint of the new notions came as early as 1900 in the case of radium D (which was not yet known as radium D). The German chemists K. A. Hofmann and E. Strauss noticed that it seemed to behave a good deal like lead. The more it was studied, the closer the resemblance grew. If it were mixed with lead, it couldn't be separated out again, no matter what. People took to calling radium D radio-lead.

And yet this was awfully disturbing. Radium D was radioactive, and lead was not. To suppose that radium D was a kind of lead was to suppose that the same element could have two sets of properties, and that was a very hard supposition to swallow at the time.

It took ten years for the inevitable conclusion to grind its way into acceptance. The American chemists Herbert N. McCoy and W. H. Ross in 1904 through 1907 pointed out the identical behaviors (chemically speaking) of certain other radioactive substances that had different properties (radioactively speaking). Still, things hung fire.

Then, in the period 1911 to 1913, Frederick Soddy advanced his theory of isotopes and described how one isotope broke down to another according to whether it emitted an alpha particle or a beta particle. An alpha particle emission meant a decrease of two in atomic number and four in atomic weight. A beta particle emission meant an increase of one in atomic number, with no change in atomic weight.

If this were so, then the lead with which all three of the radioactive series end (radium G, actinium D, and thorium D) was not merely lead, but different varieties of lead. (In those days, that sounded like saying "triangles with different numbers of sides.") Radium G ought to be lead with an atomic weight of 206; actinium D, lead with an atomic weight of 207; and thorium D, lead with an atomic weight of 208.

Now, good old ordinary lead, the kind you dig out of the ground, has an atomic weight of 207.21. Everyone knew that. It had been very carefully measured by extremely capable chemists.

You can imagine the excitement, then, when, in 1914, Harvard chemist Theodore W. Richards announced that the atomic weight of lead isolated from uranium ores was far below 207.21.

From that moment on, there was no further question. The isotope theory was accepted. It explained almost everything. For instance, radon, actinon, and thoron are all isotopes of the same element. Only one place in the periodic table need be found, not three, and one place was ready and waiting. (The element is today called radon, but some people wonder if a better name might not be *emanon*, covering all three "emanations" and not preferring one to the others.)

One later improvement that had to be made on Soddy's theory was to allow for the existence of nuclear isomers. That is, a given nucleus could be made up of a particular number of protons and neutrons that might yet take up two (or even three) different arrangements. Each arrangement would have different radioactive properties—for instance, uranium X_2 and uranium Z are the same isotope but are nuclear isomers.

Now we can summarize the phantom elements that were discovered in the first decade of the twentieth century and equate them with the isotopes of the various elements as we know them today. I refer you to the accompanying table.

Note that the isotopes of ten different elements are involved. Of these, five (uranium, thorium, bismuth, lead, and thallium) were known before 1898. Therefore, only five real elements were discovered during all that flurry: polonium and radium (by the Curies in 1898), actinium (by Debierne in 1899), radon (by Dorn in 1900) and protactinium (by Hahn and Meitner in 1917).

Furthermore, despite all the isotopes discovered, two blanks remained in that region of the periodic table. No isotopes were found of elements 85 and 87. Actually, such isotopes are formed in one or the other of the series but in such small quantities that they were not detected until the late 1930s and early 1940s.

TABLE

Element	Isotope	Phantom Element
Uranium (atomic No. 92)	uranium 238	uranium I
	uranium 235	actino-uranium
	uranium 234	uranium II
Protactinium (atomic No. 91)	protactinium 234	uranium X_2, brevium
	protactinium 231	uranium Z
		protactinium
Thorium (atomic No. 90)	thorium 234	uranium X_1
	thorium 232	thorium
	thorium 231	uranium Y
	thorium 230	ionium
	thorium 228	radiothorium
	thorium 227	radioactinium

Element	Isotope	Phantom Element
Actinium (atomic No. 89)	actinium 228 actinium 227	mesothorium 2 actinium
Radium (atomic No. 88)	radium 228 radium 226 radium 224 radium 223	mesothorium 1 radium thorium X actinium X
Radon (atomic No. 86)	radon 222 radon 220 radon 219	radium emanation, niton, radon thorium emanation, thoron actinium emanation, actinon
Polonium (atomic No. 84)	polonium 218 polonium 216 polonium 215 polonium 214 polonium 212 polonium 211 polonium 210	radium A thorium A actinium A radium C′ thorium C′ actinium C′ radium F
Bismuth (atomic No. 83)	bismuth 214 bismuth 212 bismuth 211 bismuth 210	radium C thorium C actinium C radium E
Lead (atomic No. 82)	lead 214 lead 212 lead 211 lead 210 lead 208 (stable lead 207 (stable) lead 206 (stable)	radium B thorium B actinium B radium D, radio-lead thorium D, thorium lead actinium D, actinium lead radium G
Thallium (atomic No. 81)	thallium 210 thallium 208 thallium 207	radium C″ thorium C″ actinium C″

(Also, beginning in 1939, missing elements 43 and 61 were formed by nuclear bombardment, element 93 was discovered and found to be the parent of a fourth radioactive series, and elements 94 through 105 were also manufactured.)

Having gotten that straightened out, you might think that chemists would shift, thankfully, from the old system to the new as far as naming the members of the radioactive series are concerned.

Imagine, after all, having to explain to a student that radium A and radium F are not isotopes of radium, but isotopes of polonium; that radium C and radium E are isotopes of bismuth; that radium B, radium D, and radium G are isotopes of lead; that none of them are radium at all; that radium G isn't even radioactive.

Imagine having to explain that mesothorium 2 is really an isotope of actinium, while radioactinium is really an isotope of thorium. We could drive the poor kids mad.

But alas, chemists have not seen fit to make the change. The very latest handbooks and textbooks, list the radioactive series complete with alphabet soup, assorted prefixes, primes and double primes, and so on.

In fact, when the French chemist Marguerite Perey first discovered that an isotope of element 87 was formed in the actinium series, between actinium and actinium X, she called it actinium K. Then in 1946, when she named element 87 francium, as was her right, did she say that actinium K was francium 223? No, she recommended that it still be called actinium K.

I'm not sure what moral one ought to deduce from this—perhaps that scientists, as a group, tend to be somewhat more conservative than elderly corporation lawyers.

18 · Einstein's Vision

In 1905, Albert Einstein, a junior official at the patent office at Berne, Switzerland, published scientific papers on three separate subjects. Einstein had gotten through school only with difficulty. He had tried to find a teaching position and failed. Now, at the age of twenty-six, he had no academic connection, no laboratory, no scientific instruments—only pen, paper, and his mind.

One of his papers explained what was called the "photoelectric effect." To do that he made use of the quantum theory, which had been advanced by Max Planck only five years earlier. The quantum theory had first been greeted with thorough skepticism, but Einstein's use of it made it seem so convincing that it *had* to be accepted. This revolutionized physics completely and earned Einstein a Nobel Prize sixteen years later.

Another paper worked out a mathematical analysis of something called "Brownian motion," which had first been noted by Robert Brown eighty years earlier. By using Einstein's equations it proved possible to work out the size of atoms *for the first time.*

You would think that this was enough for a young patent-office official—but also in that year, Einstein published his first papers on what came to be known as "the theory of relativity."

Einstein had begun by wondering what would happen if he were traveling at the speed of light and looked at a light wave that was traveling (naturally) at the same speed. The results seemed so paradoxical that Einstein decided it couldn't happen. To avoid the paradoxes he assumed that whenever one looked at a light wave, it would seem to be traveling at the same fixed speed, no matter how fast the observer was moving, and no matter how fast the object that gave off the light was moving. The speed of light (186,282.4 miles per second, we now know) was, in his view, one of the basic constants of the universe.

What Einstein then did was to work out, mathematically, the kind of universe that would have to exist if the speed of light was a constant. It turned out that over small distances, where nothing was moving very quickly, the universe was the one we were used to, the one Isaac Newton had worked out three centuries before.

Over vast distances, and at great speeds, however, certain differences from the usual mounted up. Lengths decreased, mass increased, the rate of passage of time slowed. It was a weird and apparently ridiculous world that Einstein described, yet it hung together mathematically—provided the speed of light was a constant. In fact, Einstein showed that if the speed of light was a constant, it was also a maximum—no object with mass could go faster.

There was no way of checking Einstein's theory in the ordinary world. Nothing went fast enough to show "relativistic effects" large enough to be detected. In 1905, however, scientists were already working with subatomic particles shot out by radioactive atoms such as those of uranium. These were indeed moving at respectable fractions of the speed of light, and when they were studied, they showed the relativistic effects exactly.

Ever since 1905, physicists working with subatomic particles have found that Einstein's theory describes subatomic events perfectly. Einstein's theory showed, for instance, that mass had to be a form of energy and that the relationship could be expressed by an equation $e=mc^2$. This meant that a very small quantity of mass could be converted into a very large quantity of energy. Physicists studying subatomic particles found this was indeed true, though no one before Einstein had suspected such a thing.

In 1945, the first atomic bomb exploded at Alamogordo, New Mexico, was a monument to Einstein's vision (though one that dismayed him), for it was designed by scientists who worked on it because they knew that $e=mc^2$.

But Einstein's first papers on the theory of relativity only dealt with certain special objects, those moving at constant velocity. It was a "special theory of relativity." Einstein tried to extend it to objects moving at changing velocities, and he finally worked out a "general theory of relativity" in 1915.

General relativity gives a grand picture of the universe, in which gravitation was explained by imagining space to be curved, with objects following the slope of the curve in their motions. The more massive and dense an object, the more sharply space curved in its neighborhood.

Light rays would have to follow the curve, too. If light from a star passed near the Sun on its way to us, that light ray would curve, and he would see the star a little farther from the Sun than it ought to be. The trouble is that any star near the Sun is blanked out by sunlight, so we can't tell. But what happens during an eclipse of the Sun?

On March 29, 1919, an eclipse of the Sun was going to take place at a time when a number of bright stars would be in the neighborhood of the Sun. The Royal Astronomical Society of London made ready two expeditions to view that eclipse. The positions of the stars near the Sun were measured. Six months later, the positions of the same stars were measured when the Sun was shining in the opposite half of the sky.

Einstein proved to be right, and the general theory was backed.

The backing was shaky, though. Whereas the special theory of relativity was supported by numerous tests in the laboratory so that its correctness is accepted by everyone, the general theory of relativity was supported, at first, only by two or three very delicate and somewhat doubtful astronomical measurements.

Ever since 1919, therefore, scientists have been seeking further tests of all kinds that would help assure them that Einstein's general theory of relativity was right—or wrong.

So far, though, Einstein's vision has held firm.

In 1958, Rudolf Mössbauer worked out a delicate way of measuring the increase in the strength of gravity as you approach closer to the center of the Earth, say, in passing from one floor in a building to the floor below. He used it to test general relativity—and the theory was supported.

In 1967, Joseph Weber finally detected gravitational waves in the universe. Their existence had been predicted by Einstein. In 1972, atomic clocks were flown around the world, and the change in their time rate proved to be what Einstein said it would be. Right now, astronomers speculate about "black holes," which are the ultimate end of matter, and those black holes fit Einstein's theory.

There may be room for improvement, but if so, we haven't found it yet. The universe, as we see it now, is, as nearly as we can tell, the same universe that Einstein envisioned in 1905 and 1915, with nothing more than pen, paper, and a single assumption to start with.

Chapter 18 • AFTERWORD

People seem to have gotten quite used to seeing me turn up in the oddest places. I myself am less inured to it. Only this month (as I write), an interview with me—a very good interview, I think—appears in *Gallery*, a magazine featuring photographs of lovely young women who had mislaid their clothes. There is a full-page photograph of myself in the magazine (a quite conventional head-and-shoulder shot), but what is on the other side of the page I absolutely refuse to tell you.

But one of the outlets in which my appearance has become rather customary, and which I myself take quite in my stride now, is *TV Guide*. For a few years I specialized in writing humorous articles for them, but then something else turned up. They took to publishing "background articles" explaining something about the subject of a TV special that might appear in a particular week, and for that they would turn to me every now and then.

I took special delight in filling a page or two of that very-high-circulation magazine with straight science. The foregoing essay is an example. So is the following one. And so are Chapters 26 and 31.

19 · The Birth of
the Bomb

As the twentieth century dawned, physicists realized there were vast energies locked into the atomic nucleus, but there seemed no way of getting at that energy profitably. Small pieces could be chipped off individual nuclei and energy would be released, but to do this, thousands of times as much energy had to be invested.

What was really needed was some system of using the energy of one chipped nucleus to chip another, which would chip another, which would chip another, and so on. It would be a falling-dominoes device, a "chain reaction." There would be the original investment of chipping the first nucleus, but that would be enough to bring about the eventual chipping of trillions.

In 1934, a physicist, Leo Szilard, saw this. A newly discovered particle, the neutron, could be absorbed by a particular atomic nucleus and would then stir it up in such a way as to cause it to produce *two* neutrons. The two neutrons might be absorbed by two nuclei, which would then produce *four* neutrons altogether. The four neutrons might be absorbed by four nuclei to produce *eight* neutrons and so on, and so on. The energy released in a tiny fraction of a second would be greater, far greater, than that of any known explosive.

Szilard, who was then living in Great Britain, was Hungarian-born, and a refugee from Adolf Hitler's paranoid racial policies. Szilard looked to the future with fear. In 1935, he patented his plan for such a nuclear explosion and assigned it to the British Government.

Actually, the device wouldn't work. It took a high-energy neutron to stir up an atomic nucleus in such a way as to make it produce two neutrons. And the neutrons produced were only low-energy, and couldn't continue the process. The dominoes stopped falling; the chain reaction fizzled.

But then came the fateful discovery of nuclear fission. It was found that the nucleus of the uranium atom broke in two when bombarded by neutrons. It wasn't just chipped; it broke in half. Nothing like this had ever been observed before.

Nuclear fission was first announced to the world in January 1939 by Lise Meitner, an Austrian physicist who had fled Hitler's Germany and was living in Sweden. A Danish physicist, Niels Bohr, brought the news to the United States even before Miss Meitner's views were published. Every physicist at once got to work on the new phenomenon.

To Szilard, now in America, this was agony. To him, the important thing about the discovery was that it required *low-energy* neutrons to bring about nuclear fission. Each fissioning nucleus produced two or three neutrons *that could continue the chain.*

By now, Hitler was deliberately pushing Europe to the brink of war, and Szilard knew that if a nuclear bomb were to be developed at all out of the chain reaction of nuclear fission, it must be under the control of the democracies. No unnecessary word must get out to Nazi Germany. He went from physicist to physicist urging voluntary secrecy and a policy of nonpublication.

There was something else. The development of the bomb would require the most delicate engineering. To do it quickly, to win the race with doom against Hitler, would take untold sums. There was only one way of getting the money, the resources, the drive that would be necessary, and that was to go to the American Government. But how could a group of hard-headed politicians be persuaded to accept the practicality of so science-fictional a device as a nuclear bomb?

Szilard discussed the matter with two other Hungarian-born

physicists who were also refugees from Hitler's tyranny, Eugene Wigner and Edward Teller. They agreed with him that the only scientist with enough prestige to be listened to by the government was Albert Einstein—still another refugee driven from his country by Hitler.

On August 2, 1939, Einstein was persuaded to write the letter, one that pointed out the feasibility of a nuclear bomb and the danger of having an enemy get it first. October 11, 1939, the letter finally made its way into the hands of President Franklin D. Roosevelt. By that time World War II had started in Europe.

Eventually, Roosevelt decided to act, and to set up a huge secret project to develop the bomb. It was a most difficult and courageous decision, for it would be a costly project, and if it failed, Roosevelt knew he would be crucified by his political enemies for wasting vast sums on silly pipe dreams. (Actually, it cost two billion dollars—not much these days.)

The project was organized under the deliberately noncommittal name of "Manhattan Engineer District," one that would give no clue to its real objective. It was better known later by the inaccurate name of "The Manhattan Project."

Roosevelt's final authorization was put through on December 6, 1941.

The project continued for several years with the greatest secrecy, employing numbers of scientists and technicians, under strict military control, in places such as Oak Ridge, Tennessee; Los Alamos, New Mexico; Hanford, Washington, and so on.

There had been so little use for uranium before this that its simplest properties weren't known precisely. Now, tons of it had to be purified, and the proper way of doing so had to be developed. Nor was it just uranium that was needed, but a rare variety, uranium 235, that had to be separated from the rest, and no one knew exactly how to do that, either.

Systems for piling enough uranium together to set up a chain reaction had to be devised. What's more, the chain reaction had to be controlled so that it didn't blow up the people experimenting with it. The first controlled nuclear chain reaction was set up at the University of Chicago on December 2, 1942. This part of the project was under the guidance of Enrico Fermi, an

Italian physicist (technically an enemy alien, in fact) who had also fled the Hitler terror.

At Alamogordo, the actual nuclear bomb was developed under the leadership of the American physicist J. Robert Oppenheimer. The first such bomb was exploded on July 16, 1945, and it was everything the physicists had expected (and feared) it would be. Szilard's vision of eleven years before was proven accurate.

Germany had by then been defeated, but two more bombs were prepared, and, on August 6 and August 8, they were used on the Japanese cities of Hiroshima and Nagasaki. That was the final straw for still-fighting Japan, and World War II came to an end.

America had won the race! And it *was* a race. German scientists were aware of the potentialities of the nuclear bomb. If they got nowhere, it was partly because Hitler's racial theories had driven dozens of top-grade scientists to the West, and partly because Hitler's intuition drove him to concentrate on rocketry instead. Soviet scientists were also in the race, working grimly on under Igor V. Kurchatov. Soviet involvement in the war, and the destruction they suffered, were so much greater than ours, however, they lagged five years behind us.

But the victory for the United States was narrower than it seemed. Suppose President Roosevelt, on Saturday, December 6, 1941, instead of giving the final authorization for the Manhattan Project (as he did), had decided to let it go till Monday.

The Sunday in between was the day of Pearl Harbor, and in the confusion that followed, it is just possible that the Manhattan Project might have been forgotten. And then, who knows, because of that one-day delay, the Soviet Union *might* have gotten the bomb first.

20 · Watch for the
Christmas Comet!

No one knows what the Star of Bethlehem was. The Gospel of St. Matthew does not explain, and we do not even know the exact year in which Jesus was born and in which the star appeared. It might have been a nova, or a conjunction of two bright planets—or it might have been a comet!

Halley's comet was in Earth's sky in 11 B.C., and that might have been the year of the first Christmas. Perhaps it was the tail of Halley's comet that pointed the way of the Wise Men toward the manger.

We don't know.

But this year* there will be another Christmas comet. On March 7, 1973, a Czech astronomer, Lubos Kohoutek, working at Hamburg Observatory in Germany, noted a dim, hazy object on a photographic plate. He studied it from day to day and found it to be a new comet, now called "comet Kohoutek," heading for the inner solar system.

Comet Kohoutek will cross Earth's orbit (rather far from us) on December 1, and will then skim around the Sun, making its closest approach—only 13 million miles compared to our own 93

* This article was written in 1973.

million—on December 28. Back outward it will shoot and cross our orbit again, closer to us. At its closest, on January 15, 1974, it will be only 75 million miles away.

The result of all this may be that Earth may be treated to a Christmas display this year that will be the sky wonder of the century. Let us see why. . . .

Far out from the Sun, astronomers believe, far beyond the orbit of Pluto, a vast belt of tiny asteroids exists. There are perhaps a couple hundred billion of these asteroids, invisible to us, slowly circling the Sun in those empty reaches of space, each one taking some tens of millions of years to make a single sweep around the Sun.

These far-distant asteroids are not like those that exist between Mars and Jupiter. The inner asteroids, close to home, are rocky in nature. The far-out asteroids, however, are largely ice.

There may be rock at the core of these far-out asteroids, but the outer regions are composed of solid water, solid methane, and solid ammonia—substances that would be vapors if the bodies were not so far out from the Sun that their temperatures remained near the absolute zero. In some cases, the asteroids may consist of a rocky gravel or even a rocky dust, cemented by the cold, cold ice.

For billions of years, these asteroids would slowly wheel their frozen way around the Sun, but every once in a long while, one might be slowed by the small pull of a distant star. Slowing, the asteroid would fall inward toward the Sun and, after many hundreds of thousands of years, would enter the planetary regions, skitter around the Sun, then go shooting out upon the long, long trail that would take it back to the far-distant asteroid belt.

Once an icy asteroid had dropped toward the inner solar system, it would remain in its new orbit, and every million years or so, it would come flashing in once more.

As it does so, the unaccustomed warmth of the Sun it approaches would vaporize its outer layers of ice and turn them into gas. The rocky dust within the ice would be suspended in that gas, so the comet would develop a kind of layer of fog, and the bright, starlike nucleus at its center (as seen from Earth) would be surrounded by a haze that would grow larger as the comet approached the Sun.

Streaming out of the Sun in all directions is a "solar wind" of subatomic particles that strikes the dust and gas about the comet, driving it outward from the Sun. The comet's very feeble gravity cannot hold that haze against the force of the solar wind.

As a result, the comet, as it approaches the Sun, develops a "tail," a length of hazy dust and gas stretching out away from the Sun, sometimes for many millions of miles. It is this tail, this exceedingly thin haze of dust and gas, that gives the comet its spectacular appearance.

The material in the tail is lost forever to the comet, and space is full of "micrometeors," minute dust particles that once were parts of comets. Every time the comet returns to the neighborhood of the Sun, it loses more of its substance, until finally only a rocky core is left, or even nothing at all.

As long as the comet only returns once in every million years or so, it takes many millions of years for it to fade off much. Sometimes, though, such a comet passes near a planet, and the planet's gravitational pull changes the comet's orbit in such a way that it shortens and remains among the planets thenceforward. It then passes the Sun every few years, and in the course of some centuries it is gone. Astronomers have actually watched comets break up, grow dimmer, and disappear as they returned over and over again.

Most of the comets we see represent these "short period" types, and all of them are faded, dying objects, with small tails, if any. The closest of all comets, Encke's comet, which approaches the Sun every 3.3 years, can only be seen by telescope, and it possesses only an almost unnoticeable haze. Only its rocky core is left.

Even Halley's comet, which approaches the Sun only every 76 years, has passed close to the Sun, and has formed and lost a new tail 25 times since the birth of Jesus. It is no longer what it was once. It is still the brightest we see regularly, but it is no longer a great spectacular.

The real spectaculars involve the occasional arrivals of the once-a-million-years comets; those that have not yet been captured by the planets; those that may never before have been seen by human eyes, and perhaps never again, either.

Such a comet may have been the one that appeared in 1843. It skimmed by the Sun at a distance of only 81,000 miles from its surface, racing from one side of the Sun to the other in only two hours. The tail it formed was 200 million miles long, and the comet was so bright that it could actually be seen in broad daylight on February 28, 1843.

Other superbright comets appeared in 1811, 1858, and 1861. The head of the comet of 1811 was over a million miles in diameter, larger than the Sun—although, of course, it was such a thin haze that all the matter in it could have been squeezed into a small mountain. At its maximum, the tail of the comet of 1861, as seen from Earth, stretched more than halfway across the sky.

There is no way, absolutely none, of predicting the approach of such superbrights until they are actually seen, and as bad luck would have it, there have been no such superbright comets seen from the Northern Hemisphere of Earth in the twentieth century. The last bright comet easily seen was in 1910, and that was only Halley's comet, which will be returning again in 1986.

But comet Kohoutek, on its way now, may prove to be a superbright this December and January so that finally our eyes will see the kind of comets our great-grandfathers saw and, for the first time, astronomers with all the highly sophisticated instruments now at their disposal, including Skylab itself, can study such a comet. For the first time, they will observe, closely, material dating back to the formation of the solar system, virtually unaffected by the energetic processes going on in the neighborhood of the Sun.

What a Christmas present to the world!

Chapter 20 • AFTERWORD

Toward the end of 1973, the world went through a comet frenzy that included me, as the foregoing essay would indicate.

Alas, comet Kohoutek didn't live up to its hopes and, at its best, was barely visible to the naked eye. Everything I said in the essay is in accord with current astronomical thinking, but it is impossible to predict just how bright a particular comet will become.

It depends, after all, on the proportion of rock to frozen gas. Comet Kohoutek was a large one, but it must have been unusually rocky, so that there wasn't enough frozen gas available to support a huge coma and tail. What a pity!

I myself went out on the *Queen Elizabeth 2* in early December 1973 to "watch for the Christmas comet." Janet was with me, and it was a honeymoon for us, for we had just been married on November 30. For four nights we watched, and for four nights there were clouds that hid the comet from view.

We were all distressed over missing the spectacle, but had the clouds disappeared, it would have been clear that there was nothing there. That might have been more distressing.

Kohoutek himself was on the cruise, a gentle, lovable man. Unfortunately, he could not quite adjust himself to the almost imperceptible rolling of the ship and had to miss one of the talks he was scheduled to give. At once, Janet volunteered me as a substitute, and I was thrown into the breach. Fortunately, I don't mind talking on short notice.

C · Life Sciences

21 · Man and Evolution

For three billion years, more or less, the evolution of species proceeded ponderously along in a hit-or-miss fashion, until (inevitably, if things went on long enough) a sufficiently intelligent species evolved. Then intelligence took a hand, and evolution was never the same again.

The key to evolution (if man is left out of consideration) is randomness.

There are mutations, for instance, small changes in the genes that take place as an animal lives, so that eggs, sperm, and young are produced that possess genes that are not quite identical to those of the parents. In any group of young of a particular pair of parents, some are a little larger, some smaller; some faster, some slower; some stronger, some weaker; some one shade of color, some another; and so on.

There is no way of telling when or in what direction these mutations take place. They are caused by such things as cosmic rays, sunlight, radioactivity, heat, chemicals—all acting more or less randomly and unpredictably.

Some mutations are more beneficial under one set of circumstances than another. An animal that runs faster will catch more food or escape enemies more efficiently than one that runs slower. The faster ones are more likely to survive. On the other hand, where an animal need not run for food or survival, slowness conserves energy, and the slower animal is more likely to survive a food shortage.

There is a steady drift, then, in the direction of more successful competition for food or for security—and there can be many directions, each with its own advantages. This drift is the result of "evolution by natural selection."

A second factor enters: changes in the environment. Because of continental drift, mountain building, earthquakes and volcanoes, wind and water erosion, and many other factors, things change. The immediate environment grows warmer or cooler, wetter or drier, and so on. These changes, too, can be considered virtually random.

In addition, the contribution to the environment by the living species that inhabit it adds a further random factor. One species evolves to become faster or stronger or less conspicuous or less tasty—and this affects other competing species that find their food supply or their security diminished (or, for that matter, enlarged).

Because of changes in environment, the changes in evolution brought about by natural selection must forever be altering in direction.

Evolution is slow, terribly slow. At least hundreds of thousands of years can be required to bring about significant evolutionary change. This means that it is not only important to consider how the environment changes, but how fast.

If large sections of the earth grow cooler, warmer, drier, or wetter very slowly, evolutionary changes keep up with it. Hair may become shaggier or sparser; water may be gathered and preserved more efficiently or less efficiently; and so on.

In order to do this, of course, there have to be mutations that just happen to carry the species in the right direction. If this just happens not to happen, the species must physically move out of the habitat into another more hospitable one, or die out and become extinct.

And if the change in the environment proceeds at too rapid a rate, then the proper mutations, even if present, cannot shift the species quickly enough. Again, the species must move or grow extinct. Seventy million years ago, some change in the environment (we don't know what) took place quickly enough to bring the last of the magnificent dinosaurs to extinction.

There are two attributes a living species can have to counter the possibility of a too-swift change in the environment. First, it may be short-lived and have dozens of generations per century. That means the mutations can build up much more quickly than in the case of long-lived species, with only two or three generations per century.

Second, species may be fecund; that is, have many young at a time. This increases the change that some among those many young will have useful mutations and survive, and that out of *their* many young, the original size of the population will quickly be restored.

This is why insects, fecund and short-lived, have so successfully evolved into all sorts of specialized niches and have developed in every direction. It is not surprising that there are more species of insects than of all other living things combined.

There are far fewer mammalian species than insect species, but among the mammals, the small rodents and bats, which are comparatively fecund and short-lived, are far more numerous in numbers and species than are the large mammals.

And then along came man. . . .

Man was intelligent enough to be a tool-designing animal in a large way, and that meant that a new kind—a much faster kind—of evolution began to take place.

Man did not have to evolve claws very slowly; he developed stone knives and hatchets very rapidly. He did not have to evolve missile projection, but developed spears and arrows. He learned how to use fire, and in this way developed an attribute that no animal had ever evolved or, apparently, ever could evolve in the ordinary way.

Over a relatively short period of time, man "evolved" through intelligence into a much fiercer and more deadly animal than he was to begin with. No other species of life could evolve quickly enough to cope with him. It meant that man began his at-first-slow, but ever more rapid climb to predominance.

The mass of humanity on earth now amounts to over 180

million tons and is still going up. It is likely that never in the history of life has a single species formed so large a fraction of all life.

In the process of increasing his numbers, man has used his intelligence to favor some species and to fight others. He has herded animals to serve as food, as amusement, as a labor force, and has fought off those predators who would compete with him for those animals. He has grown grain, vegetables, and fruit for food, destroying plant and animal life that would compete.

The result is that animals such as cattle, horses, sheep, pigs, goats, and chickens, and plants such as wheat, corn, barley, carrots, and apple trees, are much more common than they would have been without man's intelligence directing the course of evolution and increasing the security of these organisms for his own benefit. And, of course, as more and more land and food are given over to the plants and animals man desires, less and less is available for other species of plants and animals.

As man's numbers expand, then, so do the numbers of those species he favors; while the numbers of other species decline—not through the blind action of evolution, but through the weighted effect of man's intelligence. The roulette wheel of evolution has become crooked.

Not all the changes man brings about are deliberate, however, or even beneficial to himself. As man changes the balance of living species, he changes the environment, and this affects the evolutionary process, sometimes in ways he cannot counter.

When man cultivates grain in continuous large areas, animals who can feed on the grain or parasitize it find an enormously favorable (man-made) environment and multiply explosively.

Man himself, as he multiplies, offers a more plentiful and secure home to those creatures that parasitize him. Under conditions of crowding, and the necessarily poor hygiene that brings about, disease contagion becomes wild-fire rapid.

Thus, as man has used his intelligence to improve his own situation and to increase his own numbers, he has carried with him many plants, animals, and micro-organisms that he does not want, but that multiply in response to the new environment man creates.

These are dangerous in direct proportion as they are short-lived and fecund. The larger species, longer-lived and with few young, are relatively easy to counter. Even prehistoric man helped drive the large mammoth to extinction, and in more recent times, the extinction or near extinction of large animals at man's hands has become a childishly simple process.

What about small animals, though, which can hide and are difficult to find, which live on man's leavings and multiply, which are short-lived and fecund, so that even concerted and determined efforts by man can make little inroads in their numbers? They flourish. They evolve quickly enough even to counter radical changes man produces in their environment.

Thus, the introduction of insecticides can kill particular species of insects in vast numbers and reduce them to nearly the vanishing point, it would appear. A tiny fraction, however, may through natural variation be resistant to that insecticide, and these will suffice to restore full numbers in a few quick generations, and then virtually all the new strain will be resistant. In the same way, strains of germs become resistant to antibiotics.

So though the large and magnificent animals diminish in numbers from year to year, we have the rats, the bugs, the weeds, and the germs always with us.

Sometimes, mankind creates pests for himself for absolutely trivial reasons. A few rabbits, introduced into Australia, find an environment that presents them with much food and few dangers, so that they quickly multiply to fill the niche, and become pests that are fecund enough to defy efforts to exterminate them. English sparrows and starlings, introduced into the United States, are other examples.

And man's animal allies can bring about the extinction of species that, without human intervention, might have lived on for ages. Domestic animals contributed to the extinction of the dodo, and the introduction of such animals into Australia and New Zealand crowd out the native animals.

Mankind even alters the physical environment itself, as by cutting down forests, by building dams, by filling in swamps, by using the kind of agricultural practices that encourage soil erosion.

Man's animal and plant allies help, too, as when goats overgraze the land and turn grasslands into deserts, or when repeated crops of cotton or tobacco reduce the fertility of the soil.

Every alteration of the environment alters the balance of species, even when man does not deliberately intend to do so. Man may have nothing against an inoffensive forest-dweller; may even wish him well; but if the forests disappear, so must the forest-dweller. If a swamp is filled in, the plants and animals adapted to plentiful moisture find the loss of that moisture fatal. They cannot evolve away from swamp conditions as fast as mankind can reduce those swamp conditions.

Furthermore, man's crowding numbers and his industrial civilization mean that wastes of all sorts must be produced. Some of these wastes are part of the natural life-cycle and encourage the growth of particular organisms that flourish in those particular wastes—sometimes, as in algae growth in lakes, to the detriment of other species.

Other wastes are not part of the natural life-cycle and become encumbrances that do no harm and may even be useful, as when human artifacts on the sea bottom serve as anchors or shelters for sea life. But some wastes are actual poisons to which few, if any, life forms are adapted, and this is terribly dangerous.

Very few endangered species are endangered, then, because man is making any deliberate effort to reduce their numbers or to wipe them out. Most are endangered because man, in his self-centered desire to make the world more comfortable for his blindly increasing numbers, is changing the habitat to suit himself, and in the process is introducing those changes at a rate far faster than most animals and plants can evolve to meet them— although there always remains a small minority of species that just happens to be suited by man's habitat, and that therefore multiplies.

Need we do anything about it? Can we just let those species that are endangered become extinct and shrug our shoulders and say that that is just the way the evolutionary ball (as modified by man) bounces?

Not really. The world would be impoverished as life loses its diversity. Furthermore, since all species are interconnected, it is quite likely that as the ecology grows less complex we will find ourselves damaged in many ways. As a matter of fact, man's nearsighted changes of the environment are becoming self-endangering already. Man's vast numbers, dependent as they are upon rich supplies of resources, can become a catastrophic handicap if those resources dry up.

Yet if man's numbers continue to increase and if Earth's resources continue to be used without thought for the morrow, we can do nothing about endangered species (or anything else, either). The ever-more-desperate attempt to feed and care for additional billions will crowd most species (our own included, at last) to the wall in an enormous catastrophe.

If man's numbers level off and even decrease, there is a chance. If Earth's resources are used wisely, it may be that we can deliberately care for species, preserve habitats, and maintain an ecology of reasonable complexity.

Indeed, if man's knowledge continues to increase, we may attain to a third level of evolution—one that is no longer entirely hit-and-miss, and no longer even one that is directed by man, only to his short-term desires.

The third level can be man's intelligent interference with evolution, for the purpose of securing the entire ecology.

Suppose man learns how to modify genes at will, in known directions. He may then be able to introduce what he considers desirable mutations into any species, including his own, and in this way guide evolution. Mutations will be chosen for the manner in which they strengthen the ecology, make it more complex, more versatile, more stable. It may be that not only will old species be modified, but also new species, quite different from anything in the past, may be designed.

All history shows, of course, that introducing changes into the ecology will inevitably produce any number of unexpected side effects, some of which are bound to be strongly undesirable. It would therefore be necessary to learn more about ecological interrelationships so that we can predict and allow for the side effects. Very likely, the matter would be complex enough to require a very delicately programmed computer to help us in this matter. We can certainly not consult our own whims.

Evolution in the old style may then vanish altogether, and life in all its manifestations, including the human, may be designed to fit comfortably and securely into a world which, of course, will still be physically changing, but whose changes will be foreseen and provided for.

Only then will life reach a full unity and maturity, with man the guiding factor of a whole that includes him and is greater than him. We might even wonder if it is toward this end that evolution has been blindly progressing, and if perhaps this end is accomplished in only one out of a hundred, or one out of a million, life-bearing planets. And perhaps it is only in those in which that end is attained that life may then go on to the next level—whatever that next level may be.

22 · The Evolution of
Human Flight

In developing human flight, man has faithfully followed the course of the evolution of flight in the plant and animal kingdom. Man has done this not because he is a copycat, but because the evolution follows a logical progression.

Flight depends on air, for it is the air that holds an object up against the pull of gravity. None of the devices for flight on the part of any organism but man would work if air were not present.

The lighter an object is, the more easily it will be held up by air. The more surface an object presents to air, the more easily it will be held up. For an object to be both light and large of surface is to make it nearly impossible for that object to stay down. The least wind will send it flying.

A dandelion seed suspended by a feathery tuft goes drifting through the air with no effort on its own part. Since the seed itself is heavier than the tuft, it has the greater tendency to fall, and the tufted seed always rides the breeze seed-downward.

Seeds and other small objects of that size or less show no volition in flight, however. They move where the wind carries them.

The first creatures to fly with volition were the insects, which first evolved about 325 million years ago.

For the first 50 million years, insects had no wings (and there are still primitive wingless insects living today). Muscular effort hurled early insects into the air, and their light weight made it certain that they would be carried by the wind even farther than muscles alone would send them.

Some insects developed flaps on either side of the body. These flaps stabilized the direction of flight and kept the jumping insect from turning somersaults. By enlarging the surface area exposed to air without much increasing weight, they made it possible for insects to remain in the air longer and travel greater distances in a single jump. Such insects were gliders. (No gliding insects exist today; we know of them only through fossil traces.)

By 275 million years ago, the insects' gliding flaps had become movable. Tiny muscles could move them up and down as many as hundreds of times a second. These wings (as we can now call them) moved downward in such a configuration as to push against the air and lift the insect, but moved downward in such a way as to slip through the air with little resistance.

A small wing must beat very rapidly if it is to lift an object. A larger wing, presenting more surface to the air, need beat more slowly (compare the wingbeats of a housefly and a butterfly). But then, as the insect wing grows larger, it becomes more prone to physical damage. After all, it is essentially nothing more than a flat membrane with feeble supporting ribs of slightly thicker membrane.

The result is that insect wings can support only a limited weight. The heaviest known insect today is the Goliath beetle, which can be 5 inches long and weighs 3½ ounces. There are fossil dragonflies with wings 2 feet across, but the bodies those wings supported were probably only a few ounces in weight.

In order to maintain heavier objects in the air, the wings had to be made of tougher material or be made stronger with tougher internal supports.

The possibility of such wings was introduced by the vertebrates, which were generally larger than insects, and had developed stronger materials that could be used as inner stiffening.

There are, for instance, some species of fish that have enlarged fins, internally strengthened by spiny rays of cartilage. When such fish thrust themselves out of water, they can glide through the air for up to 200 yards on those fins. This is not true flying, only gliding, but the species are called "flying fish" just the same.

A number of land vertebrates can glide in the same way. Among the amphibians there are "flying frogs," which have large webs between their toes. Among the reptiles, there are "flying geckos," which have fringed membranes on the side of the body. Among the mammals, there are flying phalangers, flying lemurs, and flying squirrels. All of these have fringes of skin at the side, which fold up neatly when they are at rest, but which, when they jump with extended limbs, stretch outward and convert the creature into a flat extension of large surface area but small weight. In that way, they can glide long distances.

The gliding vertebrates are very large in comparison to insects. There are some species of flying squirrels in Southeast Asia that are three feet long, and with a weight that is measured in pounds rather than in ounces.

For true flight, however, the flaps must beat so that the animal, as each glide shows signs of petering out, is lifted to new heights to allow the glide to begin over again. With sufficiently rapid beats, height can be maintained steadily, and with proper maneuvering, animals can drop, rise, turn, accelerate, and decelerate.

But a wing, moving up and down and encountering air resistance, must be stronger than a mere stationary flap. Wings must be internally stiffened so that they can beat against the air with the power to lift a heavy vertebrate (not a tiny insect) and yet not crumple. Fortunately, mammals had available, for stiffening, something stronger than the chitin of insects or the cartilage of flying fish. They had bones.

The first vertebrates to develop extended flaps that were internally strengthened with bone and that were capable of beating were a group of reptiles called the "pterosaurs" ("wing lizards"), which developed about 150 million years ago. Their wings were adapted forelimbs and were rimmed with the successive bones of the arms and of a vastly lengthened third finger (leaving the thumb and first two fingers outside the wing).

But such beating wings require strong muscles to manipulate them, and the pterosaurs were weakly muscled and therefore were probably inefficient as fliers. They may have glided for the most part, using wing beating largely for the purpose of extending the glide. Nor were they probably very maneuverable as fliers.

Still, some of the pterosaurs were the largest of all flying creatures, past or present. The record holder is "Pteranodon," which lived about 80 million years ago, and which had a wingspan of up to 27 feet. Like the old dragonflies, though, the body borne up in the middle of all this wing was comparatively small and could not have weighed more than 40 pounds at the outside (wings and all).

From the reptiles, about 180 million years ago, there developed both mammals and birds, which maintained their internal temperature at a constant high level and which therefore had more efficient muscles and could develop large wings that could be beaten far more strongly and rapidly.

Among the mammals, only the bats developed true wings. These were stiffened by the bones of all four fingers, with only the thumb remaining inside. With wings capable of extension, contraction, beating, and bending, the bats are master maneuverers.

The best fliers, however, are the birds. Whereas bats are an exceptional flying group among the mammals, almost all birds fly. The wings are supported by fused armbones and are covered with feathers. Not only can the wing be flexed or extended, but the feathers can be spread out or brought together, with the result that the bird's wing is the most versatile flying instrument in the animal kingdom.

The birds can attain higher speeds than can any other living creature other than man. Some swifts have been measured as flying at speeds of 106 miles per hour. (The fastest-flying bat has been measured at speeds of only 32 miles per hour.)

Nevertheless, the need to support the body on air continues to place a great premium on lightness, even when wings are as efficient as those of birds. The heaviest flying bird is not likely to weigh more than 40 pounds, as did the heaviest flying reptile. (The heaviest bat weighs no more than 2 pounds.)

It is quite likely that the basic structure of animal muscle is such that no reasonable set of wings can be developed by evolutionary means that will support more than 40 pounds. At least, 3 billion years of developing life has not succeeded in accomplishing the task.

Naturally, when man first began to envy the birds and wish that he, too, could fly, he thought in terms of the structures he saw before him. He thought of constructing wings that he could attach to his arms and flap, as in the Greek myth of Daedalus and Icarus.

Unfortunately, man lacks the muscles for efficient wing-flapping and, even if he had them, he could not beat large enough wings fast enough to lift his own weight. The dream of riding a winged horse, which is far heavier than a man, is even farther out of the question; and thoughts of flying carpets and flying demons are just fantasies.

What man had to do, if he were to fly, was to make use of those principles already developed by life forms, but to do so more intensely or extremely than other life forms could, by calling to his aid materials and techniques not available to those others.

For instance, man first lifted himself off the ground by means of balloons, bags that contained gases that were less dense than air; hot air first, then hydrogen, finally helium.

The first balloons, built by the French brothers Joseph and Jacques Montgolfier, went up in 1783, and before the end of the year a man was carried aloft by one. The principle was precisely the same as the dandelion seed held aloft by the tuft. What was different was the use of hydrogen or another light gas as a lifting device, and the immense scale (compared to the seed) on which the balloon was built.

Almost simultaneously with the invention of the balloon came the invention of the parachute by the French aeronaut François Blanchard, in 1785. The device of having an extended loop of fabric offer great resistance to the air so that a dangling man descends slowly from any height is even more closely similar to the dandelion seed dangling from its tuft.

There is no volition to either a balloon or a parachute. Either moves with the wind as the dandelion seed does. To introduce some volition, the gliding principle is used. Man can use a device with flaps extending to either side to catch the wind and be supported by it in stable flight.

The flaps are not membranes but, to begin with, consisted of tough fabric supported by thin wooden rods. Because man can use materials that are much stronger than the various kinds of flaps used in the animal kingdom, he can also make the wings much larger without danger of buckling and therefore capable of supporting considerably heavier weights than the 40-pound maximum of the animal world.

Beginning in the 1870s, gliding became quite fashionable, with an outstanding early exponent the German aeronaut Otto Lilienthal.

In using a glider, man had progressed to the stage of the flying squirrel, but again on a much larger scale.

For further volition, one needed the equivalent of beating wings. Glider wings were strong as long as they were motionless. Beating them rapidly was out of the question.

But why beat them? The beating wing in birds and other flying creatures serves two purposes. It acts to catch the wind and be borne aloft—it is a supporting surface. It also beats air backward and thus pushes the bird forward—it is a driving mechanism. It isn't necessary to have a wing serve both purposes. Why not a separate device for each? Keep the wing as a support and use something else to drive.

Suppose an external source of power such as a steam engine or, better, an internal-combustion engine were used to turn a propeller. The turning propeller would cut into the air and send it backward, thus pushing the vehicle forward. The principle was used in ships, where it was water that was pushed backward. To work with air, which is a thinner medium, the propeller would have to be turned more rapidly.

Such a power-driven propeller could be added to a balloon as well, with the hydrogen (or helium) acting as supporter. This would work well if the balloon were framed into a streamlined, cigar-shaped ship, so that it might offer less resistance to the wind. The first to power such a rigid, cigar-shaped balloon was the German aeronaut Count Ferdinand von Zeppelin, in 1900.

He built a "dirigible balloon" (one that could be directed), or, more simply, a "dirigible." With the dirigible, man had a device that could stay aloft for long times and be maneuverable within limits, though only ponderously.

The dirigible was the equivalent of the pterosaur, perhaps. Like the pterosaur, the dirigibles were notable for their size, being the largest man-made flying devices ever built, as the pterosaurs were the largest natural ones. The dirigible, like the pterosaur, is extinct.

The first to power a glider were the American aeronauts Orville and Wilbur Wright, in 1903. It was then that the aeroplane was born and that man finally had a device that was equivalent to the bird.

The aeroplane has been improved over and over again since it was first constructed. It has developed the various properties of living fliers too—in some cases to an incredibly refined degree.

The power it has available, now, is enormously greater than that of birds, so that it can lift many tons of weight into the air and keep them there for hours at a time. It has instruments capable of sensing various parts of itself, its position, the properties of its surroundings so that it is as aware, after a fashion, as living fliers are. It can communicate at almost any distance by radio and radar. It even can have an intelligence of its own in the form of an autopilot.

In addition to the airplane, man also has the helicopter which, with a large propeller above, can move vertically or stand still in air. In this, man becomes equivalent to those masters of maneuverability, the hummingbirds.

The propeller pushes air backward and the plane forward. This is a form of the action-and-reaction principle. That principle can be used more directly and efficiently if the backward push is from within the plane.

The squid makes use of the direct action-and-reaction principle when it wishes to move quickly. It pushes out a jet of water and streaks in the opposite direction. Air is much lighter than water, however, and to use the same principle in air requires more power than any organism, but man, can dispose of.

Man has done it, however. During World War II, planes were developed in which the exhaust gases of burning fuel were jetted backward at high velocity through a narrow nozzle, forcing the plane forward. Such jet planes, needing no propellers, now move heavier weights faster and farther than ever planes did before.

There is one more extension of the jet principle, however, in which man has managed to pass beyond all life forms of whatever kind.

The fuel in jet planes burns in air, so that such planes can only travel if surrounded by air sufficiently dense to support the combustion. Suppose, though, that vehicles carry their own supply of air or its equivalent. With that, we have a rocketship, which no longer depends on the atmosphere.

A rocketship can jet directly upward and leave the atmosphere altogether; and, indeed, rocketships have kept men out in space for weeks at a time and have, on six different occasions, safely carried men to the Moon and back.

But all of this from beginning to end has been a natural extension of natural principles that have served as the basis for all things that fly, whether dandelion seeds or astronauts.

23 · Living Through the Winter

The second northern mission of the Kingdom of Tropicalia had returned to the capital city on the Congo River, and its leader presented his report.

"O Tropical Majesty. My men and I have visited the northern continent concerning which the first mission reported its findings 2½ years ago. Tropical Majesty, it was not as was reported. We returned to the same place, following the star charts, and there were no smiling grasslands, no green-leaved forests, no rich brown soil, no singing birds or hopping animals.

"Instead there was painful cold, and the ground was covered with a soft, white material, which turned to liquid in our hands. The liquid looked like water, but naturally we dared not drink it. There were no leaves on the trees, merely dead brown branches; there was no grass, no birds, no animals, no soil. Tropical Majesty, it is a dead land."

The King listened carefully, then rose in the awful grandeur of his ostrich-plume headdress and lion-skin tunic. Holding up his spear, he said, "Listen to my proclamation. Since both expeditions were led by honest men who tell the truth, I must believe both. The northern continent is sometimes warm and beautiful and as alive as is our own gracious land; and it is sometimes cold and dreadful and dead. Since no land can be both, but must be either one or the other, I declare the North to be bewitched, and no citizen of Tropicalia may ever visit it again."

And no citizen of Tropicalia ever did.

Of course, I made up that story. It never happened. Yet there is a true point here. The northern lands vary from warmth and life to cold and death. The trees shed their leaves as the cold approaches and the ground comes to be covered with snow. To those who have never seen such a thing happening, it must be frightening indeed. They might even think it the result of evil and deadly magic.

And dangerous it is. If the winter lasted forever, as it does in Antarctica, no living thing could endure—and Antarctica is, indeed, for the most part, a lifeless continent. Yet though the winter in the temperate zones is only temporary, three to six months of frigid cold and of dead vegetation are still dreadful. How do living things withstand the coming of cold and the dwindling of vegetation, and last through it all to the return of warmth and food?

Many animals do so by adopting a cyclic behavior that matches the cycle of the seasons, living one way in the warmth of summer and quite another way in the freeze of winter.

To do so is particularly important for cold-blooded animals— that is, for all animals but birds and mammals. The inner temperature of cold-blooded animals is usually very close to that of the outer environment. This means that in winter, their inner temperature begins to drop toward the freezing point. If the temperature drops below freezing so that ice crystals form within their cells, the animals will die.

One way of avoiding this is to lose water as the cold approaches; to dry out a bit. What water is left is bound to the complicated molecules of their tissues. Such bound water doesn't freeze as easily and doesn't form ordinary ice crystals when it does.

Some small animals such as insects may only live, as adults, through the warm weather. They leave behind eggs or larvae, however, which can dry out partially and bind what water is left. Insect eggs, cocoons, even small caterpillars can freeze during the winter, but when the warmth of the spring sun melts the ice it melts them, too, and all the living reactions within those tiny bits of surviving life speed up. The eggs hatch, the cocoons break open, the larvae begin to feed.

There are cold-blooded animals that are too large and complex to be able to do this. Exposure to actual freezing conditions would kill them.

Fortunately for them, it is the surface of the ground, and the top water layer of lakes, ponds, and rivers, that lose heat and freeze first. The deeper layers are covered by snow or ice, which are insulators. What warmth lies below is lost only slowly. This means that all through the winter, the water and mud below the ice, and the soil below the snow, remain at temperatures above freezing.

Into the depths many of the cold-blooded animals go—fish, frogs, turtles, snakes—burrowing under rocks or into mud. There they remain until the cold months of winter pass.

Why don't they starve while they're waiting?

They don't because the fires of life burn low. In general, all the reactions that go on in living tissue become slower as temperature drops. With the approach of really cold weather, however, cold-blooded animals enter a period in which the rate of the reactions (the "metabolic rate") drops particularly low and can become only a small fraction of what it normally is. The animal may breathe only once in five minutes; its heart may beat only once in thirty seconds. At this slow rate a supply of food that might ordinarily last it a day would last it two months, so it doesn't have to eat. The animal can live on its own fat and not use up that supply until winter is past.

This slow crawl to which life is reduced is found in animals that "hibernate" (from a Latin word meaning "winter") in this way. It is a way of outwaiting winter, of passing into a sort of suspended animation until the world is good again.

It is not only the cold of winter that can make it necessary for an animal to suspend activity. In some areas, the heat of summer causes lakes and rivers to dwindle during periods of drought. Creatures that live in such water must be able to survive these periods. The most extreme example of this is represented by various kinds of lungfish.

As the water begins to diminish, lungfish burrow into the bottom mud, leaving a channel to the surface. In the burrow they coil up, and their metabolic rate goes down. All the water may dry up and the mud about the lungfish may harden into clay. The lungfish, with a very occasional gulp of air, can remain alive. It is undergoing "estivation" (from a Latin word meaning "summer"). When the rains finally come and water enters the burrow and begins to soften the mud, the low flame of life begins to flicker upward in the lungfish once more. When the water is plentiful, the lungfish wriggles out of its burrow, gets rid of all the wastes it has accumulated during estivation, and becomes a swimming, feeding creature as before.

What about the warm-blooded animals—the birds and mammals?

They don't have to fear the freezing of their bodies, since they have ways of regulating their body temperature, keeping it at some constant level (98.6° F in human beings) whatever the outside temperature.

Even in the coldest winter weather, at temperatures of forty or fifty degrees below zero, they can remain warm. To do so, they must not lose body warmth too quickly to the frigid outside world, so they are insulated with thick coats of fur (or efficient layers of feathers) and with layers of fat under the skin beside.

In order to keep the temperature within high, the metabolic rate must remain high, however, since body heat arises from rapid chemical reactions. That means that a bird or mammal must constantly eat, and during the winter, getting food can be a problem.

It is not a problem for mammals and birds that live on seafood, for there is no real winter for the ocean. The water may get a little colder in the winter, and ice may even form on the surface in polar latitudes, but even so, the water beneath always stays above the freezing point. Ocean life is adapted to this condition and flourishes throughout the coldest winter. That means that seals, walruses, and polar bears can live comfortably through the winter on the Arctic ice floes, and penguins can dine well on the shores of Antarctica.

It is also possible for mammals to survive the winter by living on forms of vegetation that are to be found under the snow. Reindeer and musk-oxen can survive on a form of lichen, popularly known as "reindeer moss," which is reached by scraping the snow away from the rocks.

Some mammals, however, cut down on their food requirements by sleeping through the winter, as cold-blooded animals do. As the days of autumn decrease in length, they eat voraciously (while they still can) and grow fat. Some animals, such as chipmunks and hamsters, also store nuts and grain in their burrows.

As the mammals grow fat, they also slow down and grow sleepy. Finally, they retire to their burrows, or into caves, where the temperature does not sink too low, and sleep away the cold weather.

This is sometimes spoken of as hibernation, and most people think of the bear, for instance, as hibernating. Actually, most mammals merely sleep and do not truly hibernate. For one thing, the body temperature drops by only a few degrees, so they are still actively metabolizing. They are conserving some energy by remaining quiet, but they must rely a great deal on all the fat they have built up in their body, or on their supplies of nuts and grain. Animals that have stored away such supplies wake up every week or so to do a little eating. A bear can wake up at any time during the winter and prowl around for a while.

By the time the spring comes, these animals are still alive, but they are skinny and very hungry.

Some small mammals really hibernate, however. The smaller the mammal, the more difficult it is for it to last through cold weather. The heat of a small mammal's body is nearer the surface and more easily lost so, for its weight, the small mammal must eat considerably more food than a large one needs to, if it is to keep up its temperature.

Some mammals, therefore, such as woodchucks, hedgehogs, and dormice, give up the fight. As the cold weather approaches, they not only eat and grow fat, but their body temperature starts dropping. This places them on the road to true hibernation.

They enter their burrows, and little by little, they become almost cold-blooded, with their temperature dropping toward the freezing point. Their heartbeat slows to as little as three per minute and their metabolic rate is as small as one thirtieth of what it normally would be. Some hibernating mammals remain in this situation for over six months at a time—and survive.

Of course, not all land animals need be trapped by winter. If they can move easily and rapidly, they can follow the good weather. If they live in areas where there are summer droughts, they can migrate to other, moister areas where grass still grows, or they can move southward to stay ahead of the winter blasts.

Those animals that can move most quickly and freely are those that possess wings and that therefore can fly through the air. These can migrate far and efficiently and can survive the winter simply by leaving it. Animals that do so include some insects, some bats (which are flying mammals), and, most of all, many birds. It is for this reason that birds don't hibernate. If they can't find food during the winter, they just fly to some other place where food is plentiful.

Some birds migrate long distances to avoid the winter. Birds that spend the summer in New England may spend the winter in Florida. Birds who summer in Wyoming may winter in Texas. Hummingbirds from the United States may store enough fat in their tiny bodies to last them in a winter-avoiding flight across the Gulf of Mexico to the more tropical climate of Mexico.

The most amazing migratory feat of all is that of the Arctic tern, a gull-like, fish-eating bird that breeds and spends the summer in the northernmost parts of North America. There is no problem of food then, for the ocean teems with it.

As the summer passes, the days begin to shorten, however, so once the young birds are full-grown, all the terns take off and fly southward for eleven thousand miles (!) to the Antarctic waters, where they go through another summer of long days and ample fish. Then, when it is the turn of the Antarctic to enter the period of shortening days, the terns fly back to the Arctic again, where they lay eggs and produce a new generation of young birds.

The Arctic terns see more daylight and experience less night than any other species of living things.

We don't understand all the details concerning the ways in which animals survive the winter. We don't know exactly what makes a hibernating animal choose the time and place of its hibernation, or just what sparks the body to begin its slowdown at the beginning of winter or sparks the arousal at the end. Hormones are involved, and the spark may be partly such matters as temperature and humidity, but perhaps mostly the changing length of the day, since that is a more fixed rhythm than anything else.

Nor do we know how hibernation habits developed over the ages of evolution. Fossils give us hints of how animals changed and developed with time as far as their physical structure is concerned, but they cannot tell us anything about animal behavior. We don't know if any of the dinosaurs hibernated, for instance, or when hibernation first developed.

The migration of winged animals is even more mysterious. A particular species of bird will head southward at a particular time of year, flying along a particular route and reaching a particular winter home. Then it will fly back at a particular time along a particular route to the same area from which it started—often to the same acre, and often on just about the same day every year.

Scientists are not certain as to how birds manage to be such accurate navigators and time tellers. Some theories involve the positions of the Sun and stars, or the nature of Earth's magnetic fields, but we don't know for sure. Whatever it is, it must be quite instinctive, for even young birds who have never migrated before can make it.

What we do know, though, is that the snows of winter have not defeated life. The frost may force a retreat, but when the winter is over and the Sun shines warm again, the vegetation returns, and with that, animals who have in one way or another survived, come back again, too.

24 · The Switchboard Inside

There are some 300 million telephones in the world, and each telephone carries, in the course of one year, a thousand conversations or so, on the average. And to make all these conversations possible, it is necessary to have the ability of connecting any one of the telephones to any other. This means 45 quadrillion interconnections altogether.

Before we turn away from that final number as something too enormous to grasp, let's consider that each of us carries something still more complex crammed into his own body—crammed, chiefly, into three mere pounds of matter called a brain in the container of his skull.

The brain contains 10 billion nerve cells and 100 billion subsidiary neuroglial cells, representing nearly 400 cells for each telephone on Earth. What's more, the individual brain cell, for all its microscopic size, is far more delicately complicated than a telephone is, and is capable of far more subtle interconnections.

As a measure of brain capacity, consider memory. It is estimated that a human being can store 1 quadrillion bits of memory in the course of a lifetime. (A "bit" is the simplest unit of information, a yes or no. Is the first digit of a telephone number a 7 or not; an 8 or not?)

You will notice that all these bits of memory represent only one forty-fifth the number of two-phone interconnections in the world's telephone system, but the brain is a small object in one skull and not a planet-girdling system. Second, the brain can do far more than store memory bits: It controls a myriad factors within the body; it records sense impressions both from within the body and from the universe outside; it weighs sense impressions and dictates responses designed to best protect the body; it considers abstractions and creates new thoughts.

And a number of brains, in co-operation, have created the worldwide telephone system and much besides.

In recognizing the superiority of nature over man's inanimate productions, however, we must not be too humble. There is the time factor. The telephone system has been created from nothing in a hundred years; the human brain has reached its present perfection only after several billion years of evolutionary development.

All protoplasm has the property of irritability. It will respond to changes in the environment in some way designed to avoid damage. Even one-celled creatures will swim toward food and away from harmful chemicals.

As multicellular creatures evolved, their capacity to react appropriately depended on the continued refinement of the irritability process. Jellyfish, representative of the most simple of the multicellular creatures, have nerve cells—that is, cells in which the property of irritability has been developed to the point where that represents their chief characteristics. These cells have developed long processes (like tiny telephone wires, in a way) along which electrical impulses can reach out to the processes of neighboring nerve cells. In this way, some sensation received in one part of the body can give rise to a nerve impulse that will pass over the entire nerve network and elicit some useful response.

With the development of still more complexity, it became inefficient to have the nerve impulse travel from one nerve cell to *all* the others. A nerve cord (an organized mass of nerve cells) runs the length of the body in all animals above the jellyfish level. At periodic intervals, nerves (nerve cells and their processes) reach out to successive sections of the body. Any stimulus at any part of the body is quickly carried to the central nerve cord, where it is processed (somehow), and a message is then directed along the nerves to that specific part of the body required to make an appropriate response.

Through the hit-and-miss processes of evolution and of natural selection over the course of millions of years, nerves specialized. Some were designed to respond particularly to the impingement of light waves; others to sound vibrations; still others to certain chemicals and pressures.

Naturally, it was useful to have the specialized nerves concentrated in the forward end of the body—that portion which, when the body moved, was the first to enter a new and untried part of the environment. It came about, then, that the specialized nerves, which became sense organs, were concentrated in the head —sense organs for seeing, hearing, smelling, and tasting.

The portion of the nerve cord at the forward end, receiving all the impulses from the ever-sharpening and ever-elaborating sense organs, had to add additional nerve cells to handle, analyze, and react to it all. In this way, the beginning of a brain was developed.

One particular group of organisms developed in the direction of great elaboration of the brain and nerve cord (the "central nervous system"). Whereas most of the chief forms of life ("phyla") had a double, solid nerve cord running along the lower surface of the body, one phylum, the Chordata (to which we belong and which first arose about 500 million years ago) developed a single nerve cord, hollow, running along the upper surface. Of the Chordata, the subphylum, Vertebrata (to which we belong), protected the nerve cord by enclosing it within the cartilage or bone of a spinal column, and the brain by enclosing it in the cartilage or bone of a skull.

The chordate brain was particularly well developed in the class of Mammalia (to which we belong and which originated some 180 million years ago). Of the Mammalia, the order of Primates (to which we belong, and which originated some 70 million years ago) was the brainiest.

The Hominidae (to which we belong, and which originated about 3 million years ago), while not quite the largest of the primates, were definitely the brainiest. Homo sapiens, our own species, which has existed in its present form for at least 50,000 years, holds the record in that respect.

Elephants and whales are the only organisms to have brains larger than ours. The largest whales have brains weighing 19 pounds compared to our 3, and the largest elephant brains weigh 13 pounds. These large brains must deal with animals still larger in proportion, and would seem to be less effective than ours for that reason, among others. Dolphins, small relatives of the whales, have brains as large as ours controlling bodies no larger than ours —but dolphins have no hands, and if they are as intelligent as ourselves, that intelligence has not taken the form of manipulating the environment.

The human brain is (leaving the dolphin brain out of account) the most complex assemblage of matter known. It is far more complex than a star, for instance, so it is not surprising that scientists know more about the intimate details of a star's interior than they do of the brain.

The importance of the brain to the human scheme of things can easily be shown. We would expect the hard-working muscles of the body to be hungriest for oxygen, but they are not. The brain is. One quarter of the oxygen being consumed by the tissues of the body is used up in the brain, although that organ makes up only one fiftieth of the weight of the body and, apparently, just sits there.

And of the entire human central nervous system, that portion that is most overgrown, and therefore most humanly significant, is the cerebrum, which is the large, wrinkled portion on top of the brain. Behind and below is the cerebellum; and beneath that the medulla oblongata, which narrows to the spinal cord.

Innumerable nerve fibers bring messages to the central nervous system from every portion of the body, all of them, together, serving as a continuing monitor of the status of the internal environment, and of the nature, direction, and amount of those changes taking place in it. From the central nervous system, innumerable other nerve fibers direct changes in muscles and glands designed to preserve or to alter that status in such a way as to maximize the chances of survival and the level of comfort.

In addition, messages are received from the sense organs, which monitor the status of the outside environment. The messages issuing from the central nervous system must take into account the status of the outside environment as well.

It is dramatic to imagine that there is the equivalent of a little man, or a whole group of little men, in the central nervous system, all weighing each message as it comes in and deciding on each message going out—something like the operators in a large telephone exchange. The telephone system has given up the notion of individual manual control (since it would be impossible to hire enough employees for the purpose or to keep matters moving quickly and surely enough, even if they could be hired) in favor of steadily increasing automation—and the nervous system has done the same.

Most of the messages from within the body elicit automatic responses with no reference to the conscious will (that is, to the little men we might imagine as existing in the control room of the brain). For the most part, these automatic responses serve to keep conditions in the body steady—supplying an undeviating temperature, a steady muscle tone, a fixed blood pressure, and so on.

When change is necessary, as for instance when the sense organs apprise the central nervous system of some emergency in the outside world that requires instant, strenuous action, the "sympathetic nervous system," one of the subsystems, acts quite rapidly (without intervention of the will) to accelerate the heartbeat, dilate the pupils of the eyes and the bronchi of the lungs, inhibit feelings of pain, and slow down noncrisis-serving activities such as digestion.

The crisis past, the "parasympathetic nervous system" reverses all this.

There are also reflexes, automatic actions of muscles we ordinarily think of as amenable to the will, designed to take a quick response to a fixed local emergency. If you touch something hot, your hand will withdraw suddenly, even before you realize it is hot. The spinal cord has received the message and acted, before your cerebrum, the "realizing" portion of your brain, has.

Some responses that are entirely a matter of the will to begin with are automated with time. As an infant, you learn to walk only with the greatest difficulty, and indeed the process of balancing, of shifting the weight forward without falling, of swinging first one foot, then the other, all involve the most delicate co-operation among many muscles. It is not long, however, before the whole pattern is shifted to the automatic level, so that you are no longer conscious of any effort in walking.

In the same way, running, jumping, swimming, bicycling, typing, playing tennis, and a myriad other physical activities of varying degrees of complication can, with practice, be shifted to the automatic level, and then they become "easy." "Practice makes perfect," we say.

The skills remain difficult as soon as we try to remove them from the automatic level. If you are a fast typist, try typing while watching your fingers and see how quickly you stumble. On the other hand, imagine the risks of driving, if, through practice, your foot did not hit the brake at the first sight of an obstacle ahead before you have a chance to mull over the situation.

Yet the will can override these various levels of automation, as you know very well. You can make yourself stumble when you walk, or miss the ball when you play tennis. You are not a slave to your proficiency.

You can do this because, in playing tennis, for instance, you see, at every point, where the ball is and where your racket is. It is that constant vision ("feedback") that dictates, at every moment, your automatic response. You can use that feedback to superimpose an inappropriate response and deliberately miss the ball.

It is only when you cannot become aware of the feedback that you are helpless to control the automation of the body. You cannot adjust your own blood pressure by an act of will because you have no way of telling the level of your blood pressure at every moment. (Eastern mystics can learn to do such things after many years of practice, however.)

If you are hooked to a blood pressure device that tells you by the wavering of a needle what the level is, you can learn to control your blood pressure, making the needle go down or up, even when you don't know what it is you do, exactly, to bring the change about. And by using a device to observe the "biofeedback," you can learn in hours what it takes mystics, through meditation, years to do.

That portion of the working of the nervous system of which we are most aware, however, is that which controls our conscious, purposeful, act-of-will activities. We see an object and deliberately decide to reach for it—and do so. We see a street sign and deliberately decide to adjust the wheel of the car and make a right turn—and do so. There is no emergency driving us, no automation; we merely think and decide.

We don't know how that is done. No one has the faintest idea of what happens in the body, in the brain, in the nerve cells, that corresponds to the act of will, to the decision, to the I-guess-I'll-do-it. Still, anyone with a normal nervous system and with working muscles can do it without trouble, even though he doesn't know what he's doing.

Of course, even in the most voluntary of actions, there are automated components. If you reach for a pencil on the table, the motion of the hand is adjusted at every instant in accord with the changing distance between the hand and pencil, so that at the final instant, the hand comes to rest just as the fingers grip the pencil. We are so accustomed to judging the feedback that we are completely unaware of the fact that we are performing a most delicate and subtle action. We can get a faint feeling for the difficulty of what we so easily accomplish, however, if we watch a baby trying to reach for a pencil before he has learned to judge feedback, or a person with cerebral palsy trying to do the same with a nervous system too damaged to allow him to judge feedback.

Different parts of the cerebrum are associated with different parts of the body. There are sensory areas that receive nerve messages from particular sense organs. Messages from the eyes are apparently received in the rear portion of the cerebrum; from the ears at side portions. Messages from various internal portions of the body are received at fixed points along the top of the brain.

There is also a motor area from which nerve messages are issued to certain parts of the body, with each part the province of a different portion of the motor area. (The first hint of this was obtained when it was noted that damage to a certain restricted area of the brain, "Broca's convolution," made it impossible for a person to speak or to understand speech.)

When the sensory and motor areas are taken together, there remain large sections of the brain that seem neither to receive or send messages. It is this that has given rise to the myth that human beings "use only 20 per cent of their brains."

Not so. We might as well suppose that since most of the employees of a telephone company are not immediately engaged in receiving and sending messages, that only a small percentage of the company is being used. But we cannot ignore executives, secretaries, filing clerks, researchmen, repairmen, and all the rest.

Outside the actual receiving and sending of messages, the brain must be engaged in all the activity we have learned to call by such words as "considering," "weighing," "judging," "deciding," "reasoning," "concluding," "creating," "having insight," and dozens of others.

But how the brain does any of this, we do not know.

It is both frustrating (for the sake of present curiosity) and fascinating (for the sake of future research) that the brain, which is so intimately connected with ourselves, and so much the very essence of ourselves, is something concerning which the answer to almost every question is: No one knows!

25 · The Most Potent
Poison in the World

There are life forms which, in the course of evolution, have developed poisons designed to kill or to prevent themselves from being eaten. Venoms are produced by a variety of animals, from jellyfish to reptiles. Plants develop a variety of poisonous substances designed to taste bad to an animal that nibbles and to kill if the animal persists.

Pride of place, however, must be taken by the product of a bacterium that is to be found everywhere and that harms no one —ordinarily. It is *Clostridium botulinum*. *Clostridium* is Latin for "little spindle," which describes its shape, and *botulinum* is from the Latin word for "sausage," where it is sometimes detected.

Clostridium botulinum is a hangover from the old world of a billion years ago, when there was as yet no oxygen in the air to speak of. In those days, most life forms were simple unicellular creatures, which obtained their energy from the breakdown of organic molecules in the ocean, molecules that had been built up by the ultraviolet energy of sunlight. With the development and spread of green plant cells, the photosynthesis they carried on became more and more important. Molecular oxygen—the product of photosynthesis—filled the air, replacing carbon dioxide. An ozone layer developed in the upper atmosphere, cutting off the ultraviolet rays from the Sun.

In the new oxygen-atmosphere world, only those cells could exist that developed special enzymes to handle oxygen, for that substance was an active chemical which, if not tied down quickly, would combine with and dangerously alter the complicated chemicals of living tissue. Cells capable of utilizing oxygen had a new, rich energy source, however, and flourished mightily; the oxygen nonusers for the most part died, and vanished from the planet that for so long had been a comfortable home to them.

But they did not die out entirely. To this day, there remain on Earth bacteria that lack enzymes with which to handle oxygen and that are to be found in places where air does not penetrate. They still live on complicated chemicals, breaking them down for energy without using oxygen. To them, oxygen is not only useless, but downright poisonous. Included among these "anaerobic bacteria" is *Clostridium botulinum*, which lives in the soil and is found throughout the world in at least six varieties.

Occasionally, it encounters danger to itself in the environment. Perhaps there is a dry spell and the surrounding water films dwindle. Perhaps the soil is disturbed and the dangerous oxygen molecules percolate into its neighborhood. Under such conditions, *Clostridium botulinum* can take refuge in a device common to many bacteria.

It will develop a thick pellicle about itself and become a "spore." The pellicle protects it from the outside world. Dangerous molecules from without cannot penetrate the inner contents; precious water from within cannot leak out. The spark of life sinks low indeed, so that the spore can survive conditions that would be quickly fatal to a fully living cell. It can go without food and water almost indefinitely; it can survive temperatures near absolute zero without trouble; it can even survive up to half hour's exposure to boiling water temperatures.

Yet somehow spores, insulated as they are, can tell when the environment is favorable again. The pellicle splits, the fires of life are turned up, and the cell is active once more.

Whether actively living or in spore form, *Clostridium botulinum* is not harmful in itself to human beings. Any number of its cells or spores are blown about, with dust, into our lungs and onto much of the food we eat. Whether we take it into our lungs or our stomach, we are not bothered.

In canning food or in making preserves, however, we have to take spores into account. A quick boil would take care of any actively living bacteria that would inevitably have been included with the food, but that is not enough for spores. After sealing, cans and preserves should be patiently heated at boiling water temperatures or above for half an hour or more to make sure all spores are killed.

Spores of ordinary bacteria are no great danger, to be sure. Even if one of them found itself actively alive within the can, it could not stay alive long. The can is sealed under vacuum, there is no oxygen present, and there is no way for the ordinary bacterium to live.

Unless 'it is an anaerobe. In that case, it would need no oxygen; indeed, it would want no oxygen. A plentiful supply of complicated chemicals in the food, with no interfering air at all, is exactly made to order for it. This is the case with *Clostridium botulinum*. Any spore that survives the canning procedure comes to full life into what is, for itself, a paradise.

Within the confines of the can, the bacterium begins to produce a toxin that it does not form in the soil. This "botulinum toxin" is a large protein molecule that is easily broken down by heat. The contents of an infected can, if put through the usual cooking procedure, would probably be pretty safe.

If the contents of the can are eaten cold, however, then there is trouble, for the botulinum toxin is the most poisonous material known, at least from the standpoint of the quantity it takes to kill. Slightly less than an ounce of the toxin, properly distributed, would be enough to kill every human being on Earth.

Protein molecules are not ordinarily absorbed intact through the intestinal tract in any quantity. Almost all of them are digested and broken up first. However, occasional molecules manage to get through reasonably intact. It is for this reason that some of us suffer from allergic reactions to certain foods; and, apparently, some botulinum toxin molecules are absorbed reasonably intact, too—not many, of course, but it doesn't take many.

This absorption happens slowly, so that it usually takes twelve to thirty-six hours after eating for enough to be absorbed to make trouble.

The botulinum toxin does its work by attaching itself to the nerve endings that join various muscles. At those endings, a substance called "acetylcholine" is ordinarily produced. This helps transfer the impulse from the nerve ending to the muscle, and it is this impulse transfer that enables the muscle to contract and do its work.

The molecule of botulinum toxin somehow gets in the way of acetylcholine production. The muscle receives steadily fewer impulses and experiences a growing paralysis.

The muscles most affected are those of the eye, throat, and chest. There is double vision, as the eye muscles become unable to co-ordinate the turning of the eyeballs. Then there is difficulty in speaking and swallowing, and finally difficulty in breathing. It is the respiratory failure that usually kills.

If caught early enough, artificial lungs may be called into play; the intestines can be washed out; antitoxins may be used. At best, there is danger no matter what steps are taken, and if a comparatively large dose of toxin was ingested, probably no measures will be of any use. There have been some fifty fatalities in the United States since 1959—about a quarter of all the cases diagnosed.

Most cases of this ailment, "botulism," arise from home-made preserves where the procedure might well be slipshod. Commercial canning procedures are usually rigidly controlled and safe, but through accident or human error, there are a few cases of botulism arising from commercially produced cans every once in a while and, inevitably, deaths, too.

Botulism is a very rare disease, but it is a peculiarly nerve-wracking one since it may sit in almost any can we take off the shelf (though the odds against it are astronomic in the case of any one can). When even one case crops up, as in August 1971, for instance, the most drastic measures must be taken at once. It is necessary to find and destroy every can or jar that may have been involved in the defective procedures. Nothing less can insure continued safety against the most potent, and insidious, poison in the world.

26 · Science Is Where You Find It

In the beginning, everyone was a layman, for science in the modern sense is only four hundred years old. The early successes of science, however (Galileo and his telescope being the most glamorous example), led to its being taken up by amateur gentlemen. The best example of this was Great Britain's Royal Society, founded in 1660, which had its peak with Isaac Newton.

The character of early science, as a pursuit of scholarly gentlemen, tended to divorce science from the people and made it unfashionable for a learned man to pay any attention to what the lower classes had to say. Many unlearned views were dismissed as "old wives' tales," perhaps because old women were so insistent on the medical properties of the herbs they gathered that they made themselves a serious nuisance to the city doctors who, throughout the 1700s, believed that bleeding was a cure-all and killed more patients than the plague did.

In 1775, however, a young English doctor, William Withering, grew interested in an old wives' tale. An old lady, in Shropshire, had a secret cure she compounded out of twenty different plant ingredients, and it seemed to work. Withering managed to worm the recipe out of her and for ten years kept trying it out in various ways on his patients. He found that one ingredient did indeed have an effect, and that was the leaf of the foxglove, a very common shrub. Wonder of wonders, it did what the old lady said it did.

After ten years of testing, Withering reported his results in 1785, and the world was given the drug, digitalis, still used today as a powerful heart stimulant. Only don't go to old ladies for it; it's dispensed only by doctors' prescriptions.

And consider this: In the 1700s, smallpox was rampant in Europe, disfiguring and killing. If you recovered, however, even from a light case, you never got it again. So why not catch a light case? In Turkey, someone with a light case would hold a party for his friends and they would all scratch fluid from his blisters into their skin. Sometimes, such "inoculation" worked; a light case would result, and immunity; and sometimes, a severe case, with disfigurement or death.

Inoculation was introduced into England and into the American colonies (where Benjamin Franklin first opposed it, then supported it), but the danger was too great for it ever to become popular.

About 1771, however, an English doctor, Edward Jenner, heard from a patient that in Gloucestershire there was an old wives' tale that people who caught cowpox (a very mild disease) never got cowpox again—and never got smallpox, either.

Ridiculous, of course, but milkmaids were always getting cowpox and never getting smallpox. (That's where the legends of pretty milkmaids arose. Just having a nondisfigured face without pockmarks, in that smallpox-ridden century, was enough to make you beautiful.) Jenner thought about it and in 1796 finally gave people cowpox deliberately and then took the huge chance of trying to give them smallpox. It worked. They were immune. Thanks to old wives, Jenner had discovered "vaccination" (from the Latin word for "cow"), and smallpox was conquered.

Nor is it just medicine. People were always reporting stones that fell from the sky. The sacred stone in the Kaaba in Mecca was supposed to have fallen from the skies, and so was an object worshiped as Artemis in ancient Ephesus (mentioned in the New Testament as "Diana of the Ephesians"). Surely that was just myth.

But reports of falling stones continued and wouldn't stop. In 1807, no less a personage than Benjamin Silliman, a chemist on the Yale faculty, announced that he had seen a stone fall from the sky. President Thomas Jefferson, being told of this, said, "It is easier to believe that a Yankee professor would tell a lie than that a stone would fall from the sky."

By then, though, Jefferson was out of date. The French Academy of Science, annoyed by persistent reports from ignorant farmers concerning stones from the sky, sent investigators to put such nonsense to rest once and for all. A French scientist, Jean Baptiste Biot, in 1803, brought back the embarrassing news that the ignorant farmers were right after all. Ever since then, meteorites have been important objects of study to astronomers.

Then, too, in 1900, Hugo de Vries discovered that living organisms could give birth to young quite different from themselves. He bred these "mutations" and studied the results. Mutations are now fundamental to modern evolutionary theory.

But mutations were known to herdsmen as long as there were herdsmen. In 1791, a Massachusetts farmer named Seth Wright found that one of his sheep had given birth to a lamb with abnormally short legs. He took care of it, bred it, and developed a whole strain of short-legged sheep that couldn't jump over the low stone walls that enclosed his farm. He had made use of a mutation—but herdsmen don't write scientific reports, and learned biologists don't talk to herdsmen. So it took a century to discover what had been known all the time.

The list is long. Yoga can be ignored as Eastern mysticism, but then science discovers that the autonomic nervous system can indeed be voluntarily controlled, and now we call it "biofeedback." Killing by voodoo or curing by laying on of hands doesn't seem worthy of serious consideration until we begin to learn the power of psychosomatic influence.

Yet in defense of science, two points must be made:

First, old wives' tales merely stay stumbling attempts to understand and influence the universe, however long they endure, without *trained* study. Once scientists manage to take note of them, digitalis was purified and made truly effective; the detailed technique of cowpox vaccination was worked out and smallpox was conquered; meteorites were found and analyzed; mutations were produced and studied in detail to teach us about the inner workings of life.

Second, it is only the occasional old wives' tale that turns out to have something in it and makes the headlines. Scientists can't be expected to waste their lives following up *every* lead, since for every folk belief of value there are ten thousand that sound like nonsense because in honest truth they *are* nonsense. (No, Virginia, there *isn't* any Santa Claus at the North Pole.)

Well, then, isn't there any way of quickly picking out the one in ten thousand that is valuable and concentrating on that one? Alas, there isn't. Would that there were! If *I* knew of a way of picking out the good ones, for instance, you can bet I wouldn't be sitting here writing articles for a living.

T W O

Chapter 27 • FOREWORD

The heading "Two—"may puzzle you. It shouldn't.

You see, the title of the book is *Science Past—Science Future*. Part One is "Science Past," and it includes twenty-six essays. Part Three is "Science Future," and it includes fourteen essays. However, between the *Science Past* and the *Science Future* of the title is a dash, and Part Two is that dash. Got it?

In this "dash" part, I include two essays that are not science past, nor science future, nor science at all in any way. They are purely personal.

You may wonder why I should include personal essays in a book of this sort—but if you have read other books of mine, you will know that I am incorrigibly personal, and you won't wonder.

The first of these two essays, the one that follows immediately, was requested in October 1969, when I had just published my hundredth book. The title it was to bear was "How to Write 100 Books Without Really Trying." Unfortunately, the usual (and some unusual) editorial snags got in the way, and as publication was delayed, the title got more and more out of date.

Finally, when it was published in the spring of 1974, the title under which it appeared was "How to Write 148 Books Without Really Trying." For its appearance in this book, I will have to change the title again and, at a guess, I am making it, "How to Write 160 Books Without Really Trying."

27 · How to Write 160 Books Without Really Trying

I'm getting nervous about being introduced to strangers who turn out to have heard of me. I wait for that inevitable question, for there *is* an inevitable question attached to me. It goes as follows:

"But how do you find time to write all those books?"

It used to be 40 books when the question first started about 13 years ago, then 50, then 60. Right now, it stands at 160.*

It's not that there aren't others who are as prolific, or even more so. At least two contemporary authors have written over 400 books each. Usually, though, the literary purveyors of quantity specialize, and turn out the same sort of book in endless sausage links (no slur intended).

In my case, to the quantity is added variety. I have written science fiction, murder mysteries, and juveniles; novels and short stories; science textbooks and science popularizations and science essays; books on mathematics, mythology, history, geography, humor, and the Bible. I have written large books on Shakespeare, Milton and Byron.

So it's not just a matter of where I find time to write all those

* As of April 1975.

books, but where I find time to learn all the things about which I write.

The feeling is that there must be some trick to it, and when my answer to the question is an attempt to change the subject, my questioners get down to the nitty-gritty and begin a cross-examination.

Do I dictate my books, they want to know. (Somehow there is a general feeling that one can dictate at high speed—much faster than one can otherwise compose.) The answer is that I don't. I type everything I write. To be sure, I use an electric typewriter, out of which I can pound ninety words a minute (albeit with errors). And then, too, I type everything only twice—first draft and final copy. Not for me the perfectionism of the lapidary, who lovingly polishes each facet of his verbal gems.

Any hint that I type the final copy rouses astonishment. Don't I use a professional typist, they demand.

No, I don't. I do *all* my own typing. Nor do I have a secretary; I handle all my own correspondence, and when the phone rings, I answer it myself. Nor do I have an agent; I take care of all business details myself. And I read my own galley proofs and prepare my own indexes. I am, very literally, a one-man operation.

The questioner promptly suggests then that I have a huge, carefully itemized, and classified catalog of ideas, information, clippings, and so on.

No, I have no catalog of any kind.

Well, then, asks my questioner, more or less baffled, as the whole situation seems to grow more mysterious, am I a very rapid reader with a photographic memory?

No, not really. I read quickly but not at breakneck speed. My memory is good but far from photographic.

By now, they give up. They have failed to penetrate the puzzle, and so they say, All right, how *do* you manage to write all those books?

And I have grown puzzled myself by that time, so I generally answer "I don't know" and look baffled.

But I don't really like to be baffled, and when I was asked to write this article, I decided to try to think it through. Fortunately,

I am utterly without self-consciousness and can talk about myself freely.

But where do I begin?

There's one facet of my personality that no one asks about in the cross-examination. Perhaps they take it for granted. No one asks, "Are you highly intelligent?"

As it happens, I am, and surely that is an essential prerequisite in this task of mine of turning out a book a month on a variety of subjects and for a variety of audiences.

But what does one mean by "intelligent"? In saying that I am intelligent, I *don't* necessarily mean that I have deep wisdom, nor unusual mathematical ability, nor great linguistic talent, nor particularly sensitive insight. Nothing like that. All I mean is that I have the ability or the knack or whatever it is to score high on the intelligence tests prepared by the psychologists of our culture. The last time I took one (about seven years ago) I ended with a score of 162, and that was a pronounced underestimate. I say that because our intelligence tests are geared to score not only for "right answers" but also for "fast answers." The maximum score comes when all questions are answered correctly in the time allotted, but if one answers them all correctly in less than the time allotted (as I usually do), there is no higher score to give.

This kind of "intelligence" is important in my line of work. I can produce the information in my head quickly. My memory isn't perfect, heaven knows, but what particularly interests me I remember apparently forever, and as long as I remember it, I don't have to grope. I press the button, and there's the answer. (To be sure, I am getting on in years now and I am sorrowfully detecting the beginnings of a slow decline in my powers—but there's no use complaining; it's the common fate.)

Consider the psychological effect of this. One can be asked a question and give the correct answer instantly. Or one can be asked a question and be forced to ponder, consider several possibilities, and finally decide on the correct answer.

In both cases, one is correct. But to be correct only after considerable thought invites uncertainty and breeds the habit of deliberation. To be correct at once, and consistently, can't help but give rise to self-assurance.

And no one can write quickly without self-assurance. What does one say first? What second? How does one phrase this? How that? Is this the correct word, or should another be used? Perhaps one is on the wrong track altogether? That way madness lies; or, at the very least, a wild farrago of deletions, additions, and revisions.

For myself, the habit of remembering at once has led me on to the assumption that I am automatically right. I have come to take it for granted that I will start with the right sentence and continue with the right second sentence and the right third, and that the words I use are the correct ones, and that my thoughts are all as they should be. To be sure, this is not quite so, and when I go over my manuscript for my one and only revision, I usually make perceptible changes. The point is, though, that *while I write*, I take correctness for granted, and that keeps me from hovering anxiously over each line and from spending two hours of worried thought on every two minutes of anxious composition.

(But, says the questioner at this point, how does one cultivate the trick of instant recall? All I can say is that I don't know; I've always had it.)

Of course, one has to place the stuff inside the head first, or instant recall is useless, and there I was lucky again. I was blessed with a spindly physique as a youngster, so contact sports were not for me unless I was tired of being in one piece. In fact, a shrewd observation of my slum environment convinced me that the more I avoided the notice of my peers, the fewer bruises I would get. So I read books.

I would have preferred to listen to junk on radio if we had had one, or have watched drivel on television if that had been invented, or even have read trash in magazines if my hard-hearted father had allowed me to. As it was, a library card at the age of six was my father's notion of munificence—so I went to the library and worked my way through it.

Since I was only allowed two books at a time and could not be taken there often, I picked the heaviest books I could find, regardless of content. I remember stubbornly reading through the *Iliad* at a time when I scarcely understood a word of it. To this day I can still remember my overwhelming astonishment when I discovered that Achilles was pronounced uh-KILL-eez. I had pronounced it ATCH-illz, of course. Wouldn't you?

I read Shakespeare before we got to him in school and therefore enjoyed him. I also read voluminously and omnivorously in science and history, which not only made subsequent classwork easier but also early developed in me the habit of self-education. And self-education is, I firmly believe, the only kind of education there is. The only function of a school is to make self-education easier; failing that, it does nothing.

What's more, formal education stops; self-education never does. It was through self-education, for instance, that I was enabled to write a three-volume book on physics, although my total formal education in the subject was one year in high school; to say nothing of several books on astronomy, though my formal education in that subject was exactly nil.

And I still read omnivorously, and still borrow from the library; though now that I can afford to, I buy reference books, too, and subscribe to useful magazines. The world co-operates, moreover, by bombarding me with information. Advance copies of books are sent to me, as are publicity sheets and clippings in great quantities. Even my readers, bless them, send me their thoughts and theories.

I try to read as much as I can of all of it, and I try to answer everyone who writes but, of course, flesh and blood have their limits. Much of my mail is pure gold, and it is enough to keep me busy every evening with a backlog of reading that never, never, never dwindles.

There is one more essential point, perhaps the most essential. Even granted that I have managed to pack a great deal of stuff into my head and that I have it all on instant recall and that I have stacked up reference books for those moments when I forget-or-never-knew and that I type quickly and with assurance—I would still never write all those books if I didn't *want* to.

I've never met anyone who doesn't want to write a book at some time in the future. It seems to be the common ambition of human beings to express themselves at length, and to get their thoughts across to others. How many times have you heard some-one say, "I could write a book."? How many times have *you* said it?

Then, too, I've never met anyone who wouldn't like to have already written a book at some time in the past. Who wouldn't enjoy the feeling of holding a bound volume in his hand with his own name neatly printed on the paper jacket and on the spine and on the title page?

Between those two universal wants, that of writing a book at some time in the future, and that of having written a book at some time in the past, there exists, however, a present, in which a book is actually being written, in which paper is in a typewriter and the keys are pounding, pounding, pounding. It is *that* step that people somehow shun.

Almost nobody likes writing a book—present tense. That in-cludes professional writers.

Many a writer I know finds writing the breath of life, yet the actual process of writing a kind of agony. They produce books whose every word is the equivalent of a drop of blood, and they can go months and years in the hell of not being able to write at all.

I'm sorry for them. It may well be that when they do write, they turn out pearls, eternal gems that will live as long as man-kind does, and that their merest paragraph is worth the entire corpus of my works. If so, I am glad for humanity, but I am still sorry for them.

The point is that to cap all the lucky attributes of my life and character—spindly childhood, instant recall, and the rest—there is one more that is almost unbelievable, even to me.

I *like* the actual process of writing a book. I *like* sitting at the typewriter and watching the sentences grow.

And because I do, I can't help writing 160 books. In the first place, I find it easy to spend 70 hours a week at it (not all of it spent in actual typing, of course), because I find it fun. And at 70 hours a week, with only one revision, and no hesitations over the words and phrases, the 160 books come out without even trying—and without ulcers, high blood pressure, indigestion, constipation, or *any* of the ailments usually associated with a life consisting entirely of deadlines, one after the other.

Then, too, because I love the process of writing books, I want to do it all, and begrudge sharing any of it with typists, secretaries, and agents. And that, too, serves to increase the output.

Paradoxical? Not at all. Assistants and associates would do some of the work I ordinarily do, but the time saved would be lost again in giving them directions, asking questions, getting things wrong, and losing my temper. Not only would there be a net waste of time in the end, but also, as my production fell off and I grew morose, I would develop ulcers and a heart condition.

So I work by myself, and the sentences grow and the books turn out.

Oh, one last thing. I imagine that, in addition to everything else, one needs something called writing talent, too. However, what that might be I do not know, so there's no use talking about it.

Chapter 28 • FOREWORD

What follows is the most personal article I have written—so personal, indeed, that even I, pathologically unprivate though I am, hesitated.

However, the phenomenon of divorced parents is a common thing at present, all too common, and it occurred to a certain magazine for youngsters that they had printed many letters from daughters to divorced fathers, but had never printed one from a divorced father to a daughter.

So they asked me to supply the lack, and I asked Robyn's permission, and she said, Yes, provided she could see it first, and I wrote it in January 1974 when she was about to turn nineteen and showed it to her, and she said, All right.

And then it turned out that the magazine didn't like it. It wasn't what they were after. As nearly as I could make out, they wanted less love and more anguish. Tough! I liked it the way it was so I took it back and here it is, published *my* way.

28 · To My Daughter

Dear Robyn,

I received a letter today. It was about all sorts of things, but one sentence read: "And how is your beautiful, blond, blue-eyed daughter with the tonsils?"

I wrote back to my friend and said at once that today, on this very day, you were in the hospital having those troublesome tonsils taken out, and that I had been assured that their absence would make you more beautiful, blond, and blue-eyed than ever.

But by now you can't be surprised that your privacy is gone. If you had been doubtful about it before, you had those doubts utterly removed when you entered college last September and found that a number of the other freshmen knew all about you. After all, you called to tell me so—and in rather emphatic terms.

You're right, of course. It's because I have mentioned you so frequently in my always personal essays in magazines from *TV Guide* to *The Magazine of Fantasy and Science Fiction*. And my description of you has always taken the same form. You have always been recorded as my "beautiful, blond, blue-eyed daughter." Always. And so, you see, other people have begun to use the litany as well.

At least it's true. That you are blond and blue-eyed is a matter of fact. Anyone with eyes can see it. That you are beautiful is not a matter of fond paternal prejudice, either. No one can look at you and deny it. No one ever has.

But if my praise of your beauty is the truth, it's certainly not the whole truth.

I have seen you only half a dozen times in 3½ years since the family broke up and I left home, but it isn't your pretty face I miss. You could be as beautiful as an angel and be nothing at all if you were not much more than that.

There is, in fact, so *much* more to you, so much that gets buried under that "beautiful, blond, blue-eyed" jazz, so much that I don't talk about because I'm not sure I know how. Yet I'm writer enough to try, and if I succeed at all, it will mean another blow to your privacy, because this is an open letter.

You were (and, I'm sure, still are) fierce in the cause of right. In how many fights did you find yourself in the school corridors? Always you followed the same principle. You never started a fight, but merely waited till one existed, then counted noses, and pitched in on the weaker side. I tried to tell you that this was a sure way of getting beaten up, but you said, indignantly, "Two against one isn't *fair!*" Could I tell you that wasn't so?

I remember when you were twelve and I finally agreed to let you go down to the local shopping center for the first time without adult supervision. (It was only 0.7 miles away, but I was a suffocating father in some ways.) I remember you marching home proudly a couple of hours later with the glad tidings that you had been taken into custody by the police.

Naturally, there had to be an explanation.

It seems you had gone in the company of boys and girls your own age, and some "big kids" had taken something away from one of your "little kid" friends, and he had cried. The rest shrank back in fear, but not you. You charged the "big kids" and created enough disturbances to bring a local policeman running. You then proceeded to charge the "big kids" with the crime of grand larceny so vehemently that you were all taken to the station.

Of course, it's difficult for "big kids," however nasty, to do more than defend themselves passively against a small, pretty girl, so you didn't get battered to a pulp—but you didn't know that when you charged, did you?

You were also (and, I'm sure, still are) sane beyond your years. You were only eight when you accompanied me on my short two-block walk to the mailbox—a ritual I performed once or twice a day.

Invariably I would check to see if the letters had really gone down. It was a little insecurity dance I had.

One time when you skipped along afterward with a friend, you stood to one side and said to her in a low voice I wasn't meant to hear, "He'll go away from the box for half a block and then he'll turn and go back and jiggle the letter box again."

Your friend said, "Why?"

And you penetrated to the very core of the problem with incredible insight and said, "Because writers are crazy." I still don't know how you found that out. Surely it couldn't have been by watching me.

Then, too, you were (and, I'm sure, still are) sharp-tongued and quick with a retort. You never used it viciously, as far as I know, but you laid it on when you thought it was deserved.

I remember, for instance, how I used to suffer from your rigid views concerning my behavior. Could I do so simple a thing as sing when walking down the street? No, you were horrified at the impropriety. Could I smile at a young lady passing by and turn in order to watch her a little longer? No, you were offended to the core at the lechery. Could I put on any garment with a lively color or design? No, you were sickened at the tastelessness.

I remember once rebelling against your censorship. I put on a lovely big orange-striped pin-on bow tie that I had been aching to wear since I had secretly bought it. I then stepped into the kitchen, where you happened to be, and said, defiantly, "Well, how do I look?"

And you gave me your cool, fourteen-year-old look, and said, "Great, Daddy. Really great." Then you cocked your head a little to one side, and said, "Now if you just powdered your face and painted your nose red—"

You were constantly topping me, in fact, and after all, very few people do, if you don't mind my being a little immodest.

You were (and, I'm sure, still are) entirely honest. You were honest not because you lacked the imagination to lie, but because you lacked the kind of insecurity that led to having to lie. I'm pretty sure you were quite aware of how I felt about you, and you knew from infancy how to play on that feeling with a sure touch. So even if the truth were going to anger me, you had no fear that you might not be able to handle that.

But then, I never caught you lying to anyone else, either.

And truth-telling is a convenient habit. The only grand rule of conduct that I have ever laid down for you is that you were never to smoke cigarettes. It was one prejudice of mine that I was determined to foist on you, willy-nilly. You promised you wouldn't, and you tell me you don't—and I have no worries about it. You wouldn't and you don't.

And it's odd, but you were (and, I'm sure, still are) soft-hearted as well—even to me, where it is least necessary. You were fifteen when, as a result of teen-age whimsy, you used me as a chauffeur (a common fatherly duty, of course) just a little too carelessly. I lost time that I had wanted to apply to a little project of my own, and you *knew* that.

So I decided to feel ill used and withdrew into a shell of self-pity and wouldn't talk to you. All you had to do was not talk to me, and sooner or later, probably sooner, I would break down. I suspect that you were tempted to let that happen.

But then, after a while, I think you saw I was having trouble, and you came to where I was sitting on the couch, sat down next to me and, without saying a word, put your arm around my waist and your head on my shoulder.

It was a gentle apology, and it saved my pride.

And after I left home, you maintained an even balance and never took sides. If I have rarely seen you, I have talked to you often on the telephone, and you have been always sweet and warm and loving.

You couldn't have been my little girl forever, in any case. You were moving steadily in your own direction, long before I left. By the time you were thirteen, I couldn't walk down to the drugstore with you anymore.

"My friends might see me," you explained.

"What if they do?"

"You can't walk with your *father*," you said.

I wanted to ask, why not, but I didn't. Who was I to fight against the first soft whisper of puberty? You had simply awakened one morning and found that the world was the sort of place where fathers, even crazy ones, were simply irrelevant. I was suddenly too old for you.

Strange! You weren't too young for me.

By the time I left home you were having an active social life with any of a number of male counterparts who were *much* younger than I was, and who were fascinating in a way I could never be, even if I wore an orange-striped bow tie and painted my nose red.

So you're living your own life and would have been living your own life, with nothing but an occasional absent smile in my direction, though I had never left home.

Yet despite that, I would have put away more of the kind of memories you make possible if I had been near you. I would have enjoyed a little more all those things about you that have nothing to do with your looks and that would have made you beautiful if you were not beautiful. But since that can't be, I'm glad to have had fifteen years of it.

Robyn, you must know that it has been a pleasure having you as a daughter, and I am quite certain that, wherever you are and wherever I am, it will always continue to be a pleasure.

Your divorced father,
Isaac Asimov

THREE

Science Future

A · The Parts of Society

Chapter 29 • FOREWORD

I was asked to do this article, with this precise title, by a newly established magazine that belonged to the modern breed of women's magazines. I was, alas, suspicious.

"You don't mean something silly," I asked, "like designing a woman with breasts on the back so that dancing could be more fun?"

"No, no," they said, "be serious!"

So I was.

29 · If I Were to
Design a Woman

If I were to design a woman, how could I possibly change her from what she is at present, considering that I am a heterosexual male who has built up a set of firmly held conditioned reflexes that respond pleasurably to her exactly as she is? At least, as far as the surface, shape, and texture of her are concerned, I've learned to like what exists, and my personal vote is to have it stay.

But surface, shape, and texture aren't all there are to a woman. Before she is a woman, she is a human being, and what marks off a human being from all other species on Earth is the presence of a complex reasoning and creating brain.

Is there anything, then, that I would change about a woman's brain?

Let's consider. *If* it were true that the female brain in general were markedly inferior in some important way to the male brain, then certainly I would redesign the former. I would make it over in such a way as to have it the equal of the male brain.

I would not do this out of any abstract and idealistic longing for equality; my desires are more selfish than that. Our species faces a hostile universe, and for all its victories in the thousands of years of civilization, it now faces the most crucial decisions of all. The problems and dangers that face the human species possess within them the potential of unimaginable cataclysm, and to avoid that will surely take all the brainpower we can bring to bear on the development of possible solutions.

If the female brain were markedly inferior to the male brain, and if it could therefore not play its fair share in this great battle of mankind against itself and the universe, then I could think of no more important move for the species than to design the female brain to be the equal of the male brain. It would, at a stroke, double the brainpower of the planet, and to that extent increase our chances of survival.

But, then, is the female brain inferior? There has never been any proof to that effect. Oh, there is a great deal of talk (mostly from men) about women being more emotional, less rational, more intuitive, and less intelligent, but those are the kinds of stereotypes that suit the convenience of men who want to keep women in a subordinate position—and of women, for that matter, who are afraid to step out of a subordinate position.

When conditions are such that women must take over "men's work," they manage to do so. At every crisis, men suddenly find that they can call on Rosie the Riveter to turn out the airplanes.

With no evidence to the contrary (as opposed to self-serving opinion), and on what observation I have been able to make, I must firmly believe that the female brain is the equal of the male brain.

And yet, throughout the ages, women have been firmly kept in a subordinate position: kept from equal opportunity at any of the important work in the world, kept from equal education, kept from equal chance at leadership. So we have a world in which half the brains of humanity are artificially kept from helping in the solution of our problems, and in redesigning women, nothing can be more important than to remove whatever it is about them that helps keep them in subjection.

What can that "whatever" be? What is it about women that encourages their subjection?

Clearly, the greatest and most obvious differentiating characteristic between the sexes is that it is the women who bear the babies.

Both men and women contribute equally to the genetic equipment of the child. At the moment of fertilization, the woman's ovum and the man's sperm cell have each an equal amount of chromosome material. The ovum, however, also has a food supply to carry the developing fertilized ovum through an initial period, until some more long-term system of nourishment can be established—and that long-term system is exclusively female.

It is in the wall of the woman's uterus that the fertilized ovum is implanted, and it is from the woman's bloodstream that the developing embryo draws its nourishment for nine months and into the woman's bloodstream that it discharges its wastes those same nine months. It is the woman who must feed for two, who must grow ungainly and suffer the discomforts of that, and who must experience the pains as the baby forces its way through her birth canal.

The man's part of the task, crucial though it is, is over at ejaculation.

In a primitive society, where child mortality is high and life expectancy is low even for those who survive babyhood, the population can be maintained only by a high birth rate, and women must turn out one baby after another, not only producing them, but suckling them for months after birth, and watching them, with the love born of the pain they cost, for years after that.

And as long as a woman is a baby machine, her life must be bent to that task only. What can that life be but a subordinate one? And naturally society would work endlessly to persuade her that the role of baby machine is the only one for which she is fitted—since why else would a woman willingly undergo so intellectually stultifying a life?

Well, we are no longer living in a world where women must be baby machines. Life expectancy, at the moment, is high, and child mortality is low. Indeed, the world is crowded with human beings, and all sane people agree that, far from increasing our population still further, we should be seeking methods for humanely reducing it.

It is important for women not to have too many children, and it is certainly unthinkable for women at this point in history to have *unwanted* children.

Yet there woman is, with contraception still much more her responsibility than man's; with abortion hers to undertake; with an illegitimate child her responsibility; with an unwanted child more her responsibility than his (in fact, if not in theory), even if she is married.

The direction, then, in which I would move in redesigning a woman to fit the needs of our present society is simply stated.

I would design her so that she ovulated only at will. This would mean:

1. She would never go through a menstrual period unless, for some reason, she wanted to.

2. She could engage in sex anytime she wanted to, without any risk at all of having an unwanted baby. She most certainly could never be raped into having a baby.

3. She could have a baby (or at least a chance at one) only under conditions she felt favorable and with a man she deeply wanted as the father of the child.

Would this be shocking? Quite the reverse. It is the present situation that is shocking. To have babies blindly and uncontrollably is bestial. Every animal can do that much.

To be human, to have a brain that is able to judge and control matters through forethought, surely means having a baby when you want it and not otherwise, by means other than abandoning the joys of sex. This is what ovulation at will would mean. It would be natural, innate contraception.

I'll go further. What if a baby is conceived and there is a change of mind; what if promised favorable conditions do not develop; what if unexpected difficulties intervene? I would design a woman so that at any time during the first six months of pregnancy she could resorb the fetus at will. It would be the natural, innate potential of abortion.

I'll go still further. It is the natural ability of man to extrude his sperm cells into the body of the female and let the full load of development of the baby take place there. I would design a woman so that she, too, could, if she wished, extrude the fertilized ovum for development outside the body. She would then be no more the victim of pregnancy than a man is.

Remember, too, that an embryo developing outside the body can be more easily monitored for birth defects and, eventually, for desirable gene patterns. The evolution of the human species could then be more efficiently and (I take for granted—perhaps without justification) intelligently controlled.

If I could redesign women in this fashion, so that her role in producing the next generation is of no greater trouble to her than a man's is to him, there would be no reason whatever to keep women subordinate, and the effective brainpower of the species would double at once.

But perhaps I don't go far enough. Perhaps we ought to design women the total equal of men in every respect and develop a unisexual society that reproduces by budding, or in some other asexual fashion?

No! Sexual reproduction, by constantly shuffling the genes in every generation, speeds up evolution and increases its efficiency. And if such gene shuffling is desirable, then I prefer to keep the present system for inserting the sperm cells into the female. I would also prefer to keep some noncrucial physical differences so that a member of one sex can experience something other than himself or herself in all the play of physical love. (As I said, I have heterosexual prejudices—and *I* am the one doing the designing.)

But once the contribution of equal quantities of genetic material has taken place, once the ovum is fertilized, once the fun is done, I want all things equal thereafter.

And it's not to do women a favor, either. Let them take over half the responsibility for the salvation of the species. It won't be an easy job.

30 · Designing the Superman

Steve Austin, the hero of the TV show "The Six-Million-Dollar Man," is Superman born again.

Austin, however, hits closer to home. Superman's great talents are there only because he was born on Krypton. Since Krypton is purely mythical, none of us can be born there, and Superman can only remain a dream.

Austin, however, is a superman because a barely living bodily remnant was stitched together, and to it were added mechanical parts of great durability and power and with the capacity for delicate control. It cost (in the fictional world of the show) six million dollars.

What about real life, then? Can millionaires have themselves made into supermen? If not now, ten years from now, perhaps? And will the price be lowered to the point where someday the average junior executive, construction worker, and housewife can afford it?

It's not at all a strange thought, since it is only the culmination of what mankind has been doing for a few hundred thousand years. Improving this fragile and ultradestructible body of ours is, in fact, the name of the game we call mankind.

Every tool we have represents an improved body part. The

stone ax is an improved fist, and the stone knife is an improved fingernail. The armor of the medieval knight was an improved skin, and gunpowder is an improved biceps for throwing missiles.

These are all external to the body, though. The tools are faithless mercenaries who will work for anyone who seizes them and who will destroy today the person they were helping yesterday.

There are personal aids, designed to improve the parts of one particular person. There are spectacles to help the eyes, hearing aids to help the ear, chemicals to help the immunity mechanisms.

Such things only help established organs. They supplement but do not replace. And if the organ fails altogether, that's it. Spectacles won't help a blind man.

Still closer and more intimate is the pacemaker, which can be implanted in the heart and which can keep the ailing organic pump beating properly by the pacemaker's rhythmic electrical discharges.

But then why not an artificial heart altogether, and artificial kidneys, and artificial lungs? Why shouldn't devices of metal and plastic and polymer be made that are more durable and more reliable than the soft and precarious tissue parts that now exist within our skins?

Actually, you can build devices that will do what our various individual organs will do, but the problem is to control them. Once in the body, how do you make artificial parts work to suit one's personal convenience? How can you become aware of the light patterns an artificial eye is recording? How can you make an artificial muscle contract by a mere effort of will? How can you make a heartbeat adjust automatically to your level of activity?

There we have to call in the modern electronic capacities of science. We have to insert tiny electronic devices that can be hooked to the nerves and that can be controlled by the altering nerve impulses in such a way as to duplicate the natural controls of the natural body. It is something we can't quite do yet, but toward which scientists are working with considerable success.

A body can, in other words, control its mechanical parts by means of feedback. Their activity will adjust to the information brought in by the various sensory parts.

The study of methods for control by feedback is called "cybernetics." If a man's organs are replaced by mechanical devices that are cybernetically controlled, what we have is a "cybernetic organism" or, taking the first syllable of each word, a "cyborg."

In order to make a cyborg possible, we must depend on the brain. This is natural, since the brain is the essence of an individual. We have no difficulty in deciding that John Smith with a wooden leg or a glass eye is still John Smith. Part after part of the body could be replaced by a durable, versatile mechanism, and the person is *still* John Smith, as long as the brain is left untouched and as long as all those mechanical parts are responsive to the commands of that brain. John Smith will still feel himself to be "I," as much "I" as he ever was.

The ultimate cyborg, then, will consist of a man's brain, spinal cord, and as much of his nerves as are necessary, placed within an utterly mechanical body that it controls.

Such a cyborg can be visualized as a superman indeed, if the parts are properly designed. He can be incredibly strong by ordinary human standards, incredibly quick, incredibly versatile. As long as the brain is protected, he would be able to endure hard environments. He could explore other worlds with little in the way of life-supporting equipment. With nuclear energy for power, such a cyborg would have to supply oxygen only for the brain, and could remain in outer space far more easily than he could now.

In fact, come to think of it, the brain is a serious drawback. It *does* require protection from heat, cold, vacuum, and so on. It *must* be supplied with oxygen and glucose (and quite a bit of it, too, for the brain consumes a third as much oxygen as the rest of the body put together).

Worse yet, the brain dies. You are born with 100 billion or so brain cells, and that is your total lifetime supply. Some of them will die; in fact, some of them are constantly dying, but no new ones will be formed. Even if all other forms of death are precluded, a century of life will find you far gone on the road to senility.

Can we conquer senility someday? Perhaps, but there is as yet no hint that we will be able to do so; and there is a lot more than a hint that we can do something else—replace the brain altogether.

We are building computers that are more and more elaborate and versatile and that are more and more compact. Clearly, we will some day be able to build a computer that is as complex (or even more complex) than the human brain and that is as compact (or even more compact). There are no theoretical reasons why we can't, although there are, of course, considerable engineering difficulties in the way.

The time will come, then, when a cyborg's brain will have become useless and, instead of discarding a perfectly useful body, there will be inserted a mechanical brain that is just as good as, or better than, the organic one had been in its prime. Now the cyborg is all cyb and no org.

And if that can be done, why not make cybs to begin with?

That, perhaps, is the natural route of evolution. First, there is the hit-and-miss blindness of natural evolution, which takes billions of years to produce some species that is intelligent enough to begin a directed evolution, making use of advanced biochemical and cybernetic knowledge. The intelligent species then deliberately evolves itself into a cyborg and then into a cyb (or "robot," to use another term).

Perhaps all over the universe there are many millions of intelligent species that have evolved into cybs and that are waiting, with considerable excitement, to see if Homo sapiens, here on Earth, can manage it, too.

And then, when we have gone from org to cyborg to cyb, from man to robot, we may finally be allowed to join the great universal brotherhood of mind that (for all we know) represents the peak and acme of what life has striven for since creation.

31 · Food in the Future

In the course of the next decade, the food supply of the world will, inevitably, grow tighter.

Barring unforeseen catastrophe, the world population will increase by 700 million and will stand at over 4½ billion in 1984. That is equivalent to adding to the world a new nation with the population of China, and it is not likely that food production will expand in proportion.

Just as inevitably, the cutting edge of the food crunch will be the world protein supply. Of the various major food components, protein, on which we depend for the building of our tissues, is most essential, least common, and, therefore, most expensive.

This means that there will be great pressure to search out sources of protein in places where it cannot serve directly as food (or, at least, does not do so today) and convert it into an edible resource. The ocean must be cropped more efficiently. Protein will be obtained not only from soybeans (an increasingly important source in recent years) but also from seaweeds, algae, yeast, bacteria, and so on. There will even be increasing supplies of protein synthesized from such starting materials as petroleum and coal.

Proteins from inedible or unpalatable sources may be used to fatten animals, which can then, in turn, be eaten by human beings, but this is essentially wasteful. About nine tenths of the protein of animal feed is lost in the process of growing and fattening animals, and only one tenth of it ends on the human dinner table.

Clearly, if the protein could be fed, somehow, to human beings directly, the supply would stretch ten times as far. (This, incidentally, holds true in the case of animal food generally. Adding a link to the food chain, as when we eat plant food by converting it into animals first, always costs most of the plant food. That is why a vegetarian diet can support more human beings in a given food-producing area than an animal diet can, and why the pressure over the next decade will be toward vegetarianism the world over.)

As the years pass, then, we should hear with increasing frequency of protein meal or protein gel or protein soup. These will be additives designed to make ordinary protein-poor meals more nourishing.

This is a step away from "natural food," and will meet with some resistance from those who long for a simpler day when men grew their own grain on virgin land and shot their own deer in primeval forests. Those with virgin land and primeval forests at their disposal may do as they please, of course, but mankind, generally, as long as it does not succeed in limiting its numbers, will be forced in the direction of protein supplements, with starvation the alternative.

It might seem that the move toward protein supplements could make its first and largest advances in those lands most desperate for protein. In nations with advanced chemical technologies, such as the United States, it might well be, however, that advances equally large and startling, but far more sophisticated, will be made.

Indeed, it may well be that a new dimension in food service may open up in which we will no longer talk of protein at all.

Proteins, as found in nature, consist of large molecules, sometimes enormous ones, that are built up of long strings of smaller units called "amino acids." There are about twenty different amino acids (all belonging to the same family of chemical compounds, but different in structural detail) that go into the making of proteins. A particular protein molecule may consist of several hundred amino acids, including a number of each variety, arranged in some specific order that is always the same for that particular protein.

The number of possible arrangements of amino acids (each different arrangement producing a different protein molecule) is incredibly astronomical. It is almost inevitable, then, that the order varies from species to species and that the protein in each different living species is typical of itself and different from that in every other.

The proteins in the food we eat, then, are bound to be quite distinct from the proteins in our own body. However, when we digest protein, we break up the protein molecules in our food into separate amino acids. We absorb those amino acids and, in our tissues, put them together again in a particular order characteristic of *our* proteins.

Plants can manufacture all the different amino acids from simple molecules such as carbon dioxide, water, and minerals.

Animals cannot do this. They must find proteins ready-made in plants (or in animals that have eaten plants), pull them apart into amino acids, and put them together into the order characteristic of their own tissues. The plants, therefore, manufacture all the different amino acids from simple molecules not only for themselves, but, ultimately, for all the animals (including mankind) on earth.

There are a couple of points to make in this connection.

First, there is no magic about the intact protein molecule. There is no need to have the large molecule in the food and then break it down through digestion. That is the way it is done only because natural foods happen to have the amino acids combined into proteins. If, however, chemists break down proteins artificially into mixtures of amino acids and if an experimental animal eats no proteins at all, but only amino acid mixtures, that animal gets along just as well as if it had eaten the original protein from which those mixtures were obtained.

Second, not all amino acids are equally important in the diet of a particular species of animal. The animal body possesses the chemical machinery required to convert (within limits) one amino acid into another. This is important for the following reason. It may be that the proteins in food are skimpier in some amino acids (and richer in others) than are the proteins in the tissues of the animal eating the food. By converting the amino acids in rich supply into those in short supply, the animal can make more efficient use of the protein in its food.

Not all amino acids, however, can take part in this shuffling. Some amino acids (usually those occurring least frequently in protein molecules) cannot be formed from other amino acids in the tissues. These amino acids must be found as such in the food. Those amino acids that *must* be present in the food if the animal eating that food is to survive are called "essential amino acids."

In the case of the human being, there are eight essential amino acids, and it may be that, in forthcoming years, we will grow more familiar with their names. They are, in alphabetical order, isoleucine, leucine, lysine, methionine, phenylalanine, threonine, tryptophan, and valine.

What happens if there is a shortage of one of these essential amino acids? Suppose the protein in the food we eat contains all the amino acids in the same proportion that is found in the protein of our tissues, except that one of the essential amino acids— lysine, for instance—is present in only half the expected supply.

This means that when the body begins to rearrange the absorbed amino acids into our own tissue protein, it will run out of lysine when only half the amino acids have been arranged. Our body cannot form protein molecules without lysine, nor can it make the lysine from anything else, nor can it store the remaining amino acids and wait till some more lysine comes along. What the body does is to burn the remaining amino acids for fuel.

If an essential amino acid is in short supply, then, some of the protein is wasted. Adding more protein to the diet will be of only limited use as long as one or more essential amino acids continues in short supply.

This has been recognized for some forty years, and it is possible to fortify food with one or another of the essential amino acids (as with vitamins and minerals) and thus improve the quality of its proteins and the efficiency of its nutritive properties.

It may well be that we will move farther in this direction, however, than we have, until now, gone. If we are going to have protein meal supplements, why not break down the protein to an amino acid mixture and use that? (Will we call it AAM? Let's do so here.)

If we use natural protein to form the mix, the mixture is bound to be not quite like that in our own body, and it will be particularly serious if one or more of the essential amino acids is in short supply. However, we are quite likely to use human ingenuity to substitute for the machinery that plant cells have developed over the aeons and to be manufacturing quantities of amino acids from simple compounds. We can then add synthetic amino acids to the AAM to produce an optimum mixture.

An appropriate AAM would be the most efficient (and most digestible, it goes without saying) protein food that can exist, and in a protein-poor world it would become an essential item in the food dietary.

AAM would be particularly useful for those engaged in the mass preparation and the mass serving of food. Properly flavored or sweetened, it might make a soup, sauce, or dessert (with instructions, perhaps, to mop it up with a piece of bread to "make sure you get it all").

AAM might well be unusually desirable in hospitals, where, after surgery or after certain types of injuries or diseases, it might be necessary to accelerate healing and tissue buildup. There we might look forward to a still further refinement.

The variations in amino acid order are so enormously many and can result in such subtle changes that proteins differ in detail not only from species to species, but from individual to individual within a species. That is why we find it so difficult to transplant tissue from one human being to another. The body recognizes the transplant as "foreign" and rejects it. And the foreignness of the transplant rests almost entirely in the fact that its proteins do not quite match the proteins of the body into which it has been transplanted.

It is quite likely, then, that the precise ratio of amino acids in tissue protein differs slightly from individual to individual. This may be established, perhaps, by determining the amino acid ratio for each patient in some readily available protein, such as those in blood.

It is also possible that the details of the chemical machinery within the cell vary from patient to patient; that one patient handles a particular amino acid with less than average efficiency and therefore requires a trifle more of it in the mix. This would be more difficult to determine at once, but nitrogen-balance studies might reveal it in the course of time.

We can imagine, then, that AAM would be prepared on an individual basis and that each patient would receive a variety particularly suited to his own needs. (This adaptation of dietary needs to the precise individuality of the body would also apply to such items as vitamins and minerals, and these may be routinely added to the AAM in the individually suited quantities for complete nutrition.)

Suppose we imagine, then, a possible long-distance future in which population pressure has all but wiped out natural foods except as a basic source of AAM, sugars, fats and oils, starch, pectin, flavors, and so on. Or we might imagine an artificial environment, such as would exist in a colony on the Moon or on Mars, where a natural-food system has never been established.

In the artificial-food system we are imagining, every child, during the time it feeds on mother's milk (well, there's always that —nature's own food mix designed to meet the needs of the baby), will be thoroughly analyzed by sophisticated procedures that will determine the particular amino acid proportions it possesses and the exact manner in which its cells handle the individual amino acids. A code will be issued signifying the most efficient food mix required for that particular individual with respect not only to amino acids, but also to vitamins and minerals as well.

Shopping in the supermarket of the future, then, will be more than choosing which variety of simulated food you want. (No doubt, objects will be produced that in texture, appearance, and flavor will resemble the various items we are now used to.) Each individual will punch his own body code into a console, and a computer will then see to it that his order is formed out of a mixture designed to be most efficient for his body.

Of course, if mankind learns to limit its numbers, it may never come to this.

32 · The Amusement Park
of the Future

The whole notion of parks for public use is modern. In ancient times only the aristocracy could afford to have tracts of natural greenery for their own use. Of course, cities were small at that time, and those who wished to do so could easily reach open country.

In the nineteenth century, though, cities had grown so large that people within them could find no place for themselves beyond the concrete. What's more, with the growth of democratic ideas, it began to seem that there ought to be gardens and rolling fields for the general public, and not just for the rich and titled.

It isn't surprising then that the nineteenth century saw public parks designed first in those nations where democracy was best established—in Great Britain and the United States. The first large city park designed for public use in America was New York City's Central Park, which was established in 1850.

At first parks were thought of as merely patches of unspoiled greenery—walks, lawns, trees, flower beds, zoos, and botanical gardens, all designed for passive enjoyment. People were supposed to stroll through parks, free for a moment from the noise and hurly-burly of the city streets.

As time went on, though, it came to be felt that there was

more to leisure time than relief from noise. Why not a chance for active play? Ball fields and playgrounds were therefore established in parks, as well as opportunities for boating, for listening to concerts, and so on.

Other amusements far older than the public park were the entertainments that were associated with trade fairs, with exhibitions, with religious celebrations—with anything that brought many people together. Clowns and acrobats could make money by amusing the bored crowds. Others might tell fortunes, display their strength, or charge people to play games of chance.

As time passed on, society grew more machine minded, and special rides were developed that usually gave people the thrill of falling without being hurt. The roller coaster is the best-known device of this sort. The Ferris wheel was invented in 1893 and was used at the World's Columbian Exposition in Chicago that year; the parachute jump was invented still later.

Amusement areas of this kind were finally associated with parks or beaches and formed permanent establishments. There can be few people in the United States who haven't visited some amusement park at some time and ridden the roller coaster and driven electric cars that bump each other, and gone through dark tunnels of love, and seen freak shows, and tried their skill at shooting and throwing baseballs, and consumed tons of hot dogs, soda pop, and cotton candy. There are still many people who remember the great days of Coney Island and such amusement centers as the Steeplechase and Luna Park. Millions of people today have visited the great modern amusement park of Disneyland.

But what will it all be like in the future?

Of course, many things won't change. Earth's gravity is still the greatest thrill producer, and it's hard to improve on a roller coaster for scaring you out of your wits without hurting you.

There might be additions, though, that will reflect scientific advances. For instance, there might be new games that simulate outer-space maneuvering. You might have a chance to sit at a control panel and direct weak laser beams at enemy spaceships, with a bull's-eye hit seeming to blast them out of the sky.

The spaceships might be simulated on a screen, and a computer might control them. In fact, beating a computer at any of a hundred different games might be a big thing at an amusement park of the future.

Again, holography might be added to the more ordinary snapshot booth. Instead of the picture of yourself automatically taken and developed, you will get a cloudy, featureless film, which you can then have projected into a very lifelike three-dimensional image.

The time may come, too, when robots will be sufficiently advanced to take part in the pleasures. They can exhibit feats of strength, or display their mathematical powers, or (in more humble fashion) serve as ushers or baby-sitters. A piggy-back ride on a robot's shoulder might be the most popular item for younger children in the amusement park of the future.

The most important changes in amusement parks, however, can't be surely predicted, because they will depend on the kind of society we will develop. We can easily imagine some tragic changes that would break down our complex society. In that case, there might be no amusement park at all in the future, and very little of any amusement at all.

Or else there might be great changes that would not be particularly tragic but would make future life very different from the present.

Suppose, for instance, that a time comes when the world's population has dropped and become stable, and that the cities of mankind are built more and more underground. There would be certain advantages to this, for an underground city wouldn't have to worry about weather and about temperature change. Hardly any energy would have to be used to heat it and cool it. Then, too, the whole world, being without night and day, could run according to a single, staggered time schedule.

And since man and his works would be underground, there might be ways of lessening the effect of mankind on the surface ecology. Perhaps the surface cities would be gradually dismantled and used as sources of raw material with which to build the underground world. The surface, relatively free of human occupation, would then all be given over to farmland, to parks, to wilderness.

In such a world man might actually be closer to nature. Nowadays, someone living deep in a city has to travel miles in jammed conveyances through crowded streets to reach open land. In the underground city of the future, it would be just a short elevator ride upward to a vast parkland.

In an underground world, there would probably be underground amusement areas which, in many ways, would resemble those we know today. However, amusement areas have been associated with parks and beaches for so long that an underground city would be sure to establish amusement areas on the surface for the public.

Such surface amusement parks, however much they might be like present-day examples, might well have one item that we ourselves would think of as perfectly ordinary, but that may be the most popular of all rides among the people from underground. It would be a monorail perhaps, or even an old-fashioned set of connected trolley cars.

Such a ride would have no ups and downs, no sudden banks, no frightening features whatever. It would merely represent a half hour of smooth progress in the open, with trees about, with a view of fields, or stands of corn, or herds of cattle. It might be called "Ride with Nature."

The amusement park itself might well be equipped with a transparent dome that automatically darkens as the sunlight brightens, and with artificial lights that brighten as the evening darkens. Between that and the air conditioning, the merrymakers from below would have the steady weather conditions they are used to.

But many, especially the younger ones, won't want that. To feel natural weather about them would be one of the excitements of the surface park. On the "Ride with Nature" there would be the chance of feeling the wind, or the hot Sun, or seeing what the world looks like when rain is falling. The amusement park might even seed clouds under conditions that would produce light, warm rains, and then provide plastic coveralls for those who would want to experience a natural shower.

Think of what it would be like in a northern winter, when

parties of young people, dressed warmly against a cold they never experience underground, fill the elevators to the amusement parks so that they can take the ride over unbroken fields of snow and enjoy a world in pure white.

But maybe the best show of all would come on a clear, mild night, when the merrymakers climb to the upper observatory platforms. Then the lights in the park might dim down to nothing and the transparent dome would pull back so that there would be nothing but the sky overhead—and everyone would just look at the stars.

Or suppose the time comes when mankind has developed space exploration to the point where there is a working colony on the Moon. Might we not then have amusement parks on the Moon, as well as laboratories, observatories, and factories?

If so, tourists from Earth will have a chance to indulge in amusements absolutely impossible on the home planet, quite apart from the pleasure of just looking at the moonscapes and at the exciting sight of the Earth in the sky.

The Moon's gravitational pull is only one sixth that of Earth, so that every motion on our satellite would have a different feel to it. You will be able to jump higher and farther, and you will go up for a longer time and then take an equally longer time to come down. Learning to handle your body under lower gravity would take time, of course, and you will surely have to take lessons. Not many should be required to learn how to run in long, loping steps without falling.

Going down a slope would be even more exciting. Because you weigh less, you press less hard against the ground and there is less friction. That means that the Moon's surface is more slippery than the same sort of surface would be on Earth. If a crater slope is smoothed out and if you wear special spaceboots with polished soles, you could go sliding down one as though it were a ski slope on Earth.

But it would be much better. Without the feel of rushing air past your face and body, the motion would seem so effortless that you would feel as though you were floating on nothingness and it was the moonscape that was moving past you.

Of course, motion on the surface would be hampered by the spacesuits you would have to wear, but the real amusement parks on the Moon would be underground, where no spacesuits would be necessary.

The gravitational pull is just as low under the Moon's surface as on it, and you can move along springily, eat some of the lunar delicacies. On the Moon, there would be little chance for wheat-fields and chicken farms, and food imported from Earth would be too expensive for sale at amusement parks. But there would be the various microscopic forms of life grown by the colonists under artificial light. Algae and fungi grow very quickly and contain almost no wastes.

Many mutations would be cultivated and selected for flavor and texture. Skillfully they would be molded, shaped, colored, and flavored. You might suck at a frozen yeastsickle, or eat piping hot algaeburgers, part of which had the consistency and flavor of bread and part of meat. There might be tastes not quite like anything you've ever experienced, and you may well find yourself smacking your lips over them—and then you'll have to work off some of the weight you gain on the more active items of amusement.

One popular device, for instance, might be a large and complicated jungle-gym. Under the low gravity, you would be moving horizontally no more slowly than on Earth, but you would be dropping downward much less quickly. You would be able, therefore, to travel longer horizontal distances from bar to bar. You would also be able to pull yourself upward with less effort than on Earth.

With enough practice, in fact, you could go swinging about the bars like a gibbon.

And imagine what a trampoline would be like on the Moon. Since you would remain in the air six times as long for a given jump, you would be able to indulge in somersaults and side twists of a kind impossible on Earth. That, too, would be almost like flying.

In fact, why not fly? On Earth we can't fly by flapping our arms, even if we attach wings to them. Earth's gravity is too strong, and the density of Earth's air isn't great enough to support us against that.

On the Moon, however, the gravity is so low that if we imagine a large chamber filled with air under somewhat greater pressure than on Earth, personal flight without engine power might well be possible. We can imagine tough, flimsy sheets of wire-supported plastic attached to each arm. We can sail off into the air from a point high on the wall, using the plastic wings as gliders. Perhaps, once we grow accustomed to the sport, we might even be able to flap them properly in order to gain altitude.

Such a flying-chamber would probably be the most popular ride on the Moon, and Earth youngsters (whose lightness would make flying easier for them than for grownups) might become quite adept at banking, veering, and gaining altitude. They might stay in the air for quite a while before gliding down to the floor of the chamber at last.

The closest thing to that on Earth is riding a glider, but even that isn't quite the same thing. It doesn't duplicate the feel of having your own arms control and guide your flight.

The ultimate amusement park, however, will be an amusement world; a small, spherical world, with no gravity at all.

We are already in the process of building space stations that will orbit the Earth in free flight. Right now, we are thinking of them only in connection with scientific research and as a base for launching spaceships out into space.

Eventually, though, why not a large space station designed for amusement and recreation? In an age in which people can take a trip to the Moon without too much trouble and expense, it will be easier still to visit "Spaceland," a mile-wide space station orbiting Earth from pole to pole.

From "Spaceland" you can look down upon an Earth that will fill most of the sky. You will watch its cloud patterns; see its storms; catch glimpses of ocean and desert and icecaps. At times, perhaps, there will come the glimmer of a city on Earth's night side (unless all cities are underground by then).

There will be the excitement of learning to live under zero-gravity conditions. A bed might be no more than a soft sheet suspended in midair by poles at its four corners. You might have to fasten a second sheet over you or get into velvet shoulder-straps so that you don't puff off your bed with a snore or whirl off when you turn.

Eating would be a special adventure, too, especially when you have to handle liquids. It might be fun to try to make the soup or the milk form a quivering sphere in midair and then drink it through a straw or suck it up through pursed lips. However, if an air current or a careless move on your part allows the liquid to make contact with you or your clothes, you will get just as wet as on Earth.

Then think of the ball games. Imagine a spherical room with its walls well padded. Within that room, you can fly even without wings. You can jump up anywhere from its wall (wherever you stand will seem like a curved patch of floor to you) and can then maneuver your arms and legs to guide your flight. (Flippers on your feet and arms might help.)

Naturally, you would be clumsy at first, but with practice there would come skill. Imagine several of you in the air at once and one ball that must be maneuvered into one of two pockets in the wall. It would be a three-dimensional, no-gravity melee, all of you trying to maneuver to get at the ball—which may well get away from all of you. Naturally, you would all have to wear padded clothes, helmets, and mouthguards or there might be some damage in the excitement.

Or there might be a net across the room, with the two teams on either side and with no passing the net allowed. Only the ball goes back and forth.

Or perhaps once the teams have become really expert at maneuvering under zero gravity, each player can wear some kind of weak-reaction motor so that he can zoom about more rapidly and with better aim.

The possibilities are endless, and no one who will have played zero-gravity ball on "Spaceland" will ever think much of the simple two-dimensional ball games on the gravity-bound planet of Earth.

33 · Sex in Space

Eventually, if our space effort continues, we will establish a colony on the Moon, and possibly on other worlds, too. Eventually, we will engage in long manned space flights to Mars, or even farther.

This means that human beings will, for the first time in history, be away from the planet Earth for extended periods of time —months certainly, perhaps years. Eventually, if we look far enough ahead—lifetimes.

If this is so, it seems quite inevitable that there will be sexual activity, also for the first time in history, away from the planet Earth.

Will sex be any different out there?

At first thought, we might say: Why should it be? In any spaceship and in any other-world colony, mankind will do its best to create an earthlike environment. An earthlike atmosphere and an earthlike temperature will fill the ships and the colonies. Men will stay men and women will stay women. What difference can there be, you might therefore argue, as far as sex is concerned?

There is one fact, however, that mankind cannot alter very easily, but must take more or less as he finds it. That is the gravitational pull in non-Earth situations.

The surface gravity on the Moon is only one sixth that on the Earth. The surface gravity on Mars is only two fifths that on the

Earth. In coasting spaceships, the gravitational effect is just about zero.

If a ship's rocket engines are working, or if the ship is spinning rapidly, there will be acceleration effects that will feel like a gravitational pull. These will be either weak or temporary or both, however.

It is doubtful that any colonies will ever be established where the gravitational pull is *greater* than that on Earth. In the entire solar system, only the Sun and the giant planets have high gravities, and we probably won't go anywhere near them.

In space, then, whether on spaceships or in space colonies, we will be subjected to low gravity, even to zero gravity, and that has to be taken into account.

Will lack of a gravitational pull prevent sex?

Certainly not, since gravitational pull is in no way essential. To be sure, the weight of an individual on top will keep him firmly pressed against the one below. This close and extensive contact is pleasurable, but it must be carefully regulated. A man on top, for instance, must generally labor to prevent his full weight from pressing against the woman underneath, and, with the best will in the world, it is sometimes necessary to alter the position to allow her to catch her breath.

Under lower gravity, the discomfort of too much weight decreases and vanishes, but the pleasure of close and extensive contact need not. Even under zero gravity, the grip of arms and legs will be quite sufficient to assure all the contact desired.

In fact, under zero gravity, contact would be improved, for then intercourse can take place in midair and there is no necessity to be aware of the feel or pressure of anything but your partner. There need be no mattress, no bedclothes, no pillow—just one another.

Furthermore, without any gravitational pull there is no definite impression of "up" and "down" forced on you by the outside world. You can imagine yourself on top of your partner, or underneath, whichever you please, and you will probably experience the effect you concentrate on.

(To see what I mean, lie in bed on your back with your eyes

closed and without touching either headboard or footboard. Nothing in the outside world forces you to know which way you are facing. You can imagine yourself to be with your head toward the headboard, or with your head toward the footboard, and each way will feel just as natural and real to you.)

In zero gravity, you can imagine yourself to be now above, and now below, as you choose. In fact, both partners can imagine independently. At zero gravity, *both* can be on top, or both below. For that matter, you can imagine yourself vertical while your partner is imagining a horizontal position.

This doesn't mean that there aren't difficulties involved in zero gravity. You can't just jump up and hang in midair. If you're a novice, you will just travel from one end of the room to the other. Jump, and you'll simply reach the other end of the room. Jump again, and you'll be back where you started. If you jump slowly, you will move more slowly, but you will still reach the other end of the room.

In a vacuum, there would be nothing you could do just by jumping, but inside a ship there will be air present, and air has resistance. If you jump *very* slowly, air resistance may be enough to stop you halfway.

The thing is, then, to increase air resistance. Someone used to zero gravity, will have learned to move his arms in such a way as to increase air resistance when he wants to stop. In effect, he would learn to "swim" in air, making himself go faster or slower, or turning, by shifting and moving his arms and legs to take advantage of air currents and air resistance. Light fins on arms and legs would make such "swimming" easier and more efficient.

A particularly interesting way to come to a halt in midair is to involve your partner and make the attempt to come to rest in the middle of the air a part of the sex play itself.

Suppose you are standing on the floor of a large room and your partner is standing on what to you is the ceiling. You both leap upward toward each other and catch each other midway. The two opposite motions are largely canceled. You are now together and more or less motionless in midair. If there is some left-over motion, or spin, it can easily be canceled out by the proper "swimming" motions.

In fact, two partners, each of whom is a skillful air swimmer at zero gravity, can make of sex an active game indeed. Pursuit and avoidance become three-dimensional play, of which the ordinary fun and games on a two-dimensional surface are but feeble imitations.

Without concern for weight, moreover, all sorts of positions become possible that would be most unstable on Earth's surface. A person could rest comfortably on another person with any part of the first body making contact with any part of the second person, without any strain whatever.

One body can slip smoothly along another body, without the necessity of crawling or inching along a floor or bed. There would be a kind of poetry of motion that would make sex far more like a dance and less like digging ditches—and without any diminution of pleasure.

It may sound as though all this can be duplicated in water, that sexual play while floating in or under water is rather like what I am describing sex at zero gravity to be.

That, however, is wrong. It is true that in water most, even almost all, the gravitational effect is neutralized. Still, we cannot breathe under water, so that to duplicate the freedom involved in zero gravity, we must imagine the two partners equipped with oxygen cylinders and mouthpieces. That would certainly make it difficult to use the mouth for anything else.

Even if we disregard this, water is viscous. That means it is hard to force your way through water. Everything must take place in a kind of slow motion that is frustrating and energy-consuming.

Air, on the other hand, has low viscosity. You can move through it rapidly. Whether in foreplay or in the sex act itself, you can move as quickly as you wish.

Weightlessness requires precautions, of course. There is no gravitational field pulling at you, so you feel very light. You still have all your mass, however. That means that it is just as hard to start yourself moving or to stop yourself moving as on Earth. It is just as hard to change directions.

If you and your partner are whirling and your grip breaks, you will both go flying in opposite directions. There will be no gravitational pull to make you drop faster and faster. You can brake your fall by "swimming." Just the same, if you hit the wall at a certain speed, you will smash against it with all the force of hitting the same wall on Earth at the same speed.

People skilled at maneuvering at zero gravity wouldn't let this happen very often, but there is no use in taking chances. The large spaceship of the future may contain special rooms devoted to sex. They would be rather large to allow plenty of maneuvering. They would probably be spherical in shape, since there is no point in having distinct walls, ceilings, or floors—these would have no meaning at zero gravity. There would be no need for furniture of any kind; in fact, if present, they would probably represent dangers.

The spherical surface would be padded with some washable material, and this might become part of the play in itself. The room might be one huge trampoline.

Even so, complete safety is impossible. If, in the excitement of foreplay, a flailing arm meets a face, the result can be a broken jaw, zero gravity notwithstanding. The feeling of weightlessness is bound to give one too much of a feeling of freedom. The best lover will be the one who never lets that freedom become too seductive, and always remembers that although he may feel like a feather, he isn't really a feather.

Sex at low-but-not-zero gravity, as on the surface of Mars or of the Moon, can never be quite as free as in the spherical zero-gravity love room of a large spaceship. It will be more free, though, than sex on Earth.

If we imagine a colony under the surface of the Moon, with large, air-filled rooms, we cannot expect love-making in midair. There is enough gravity, even if that is only one sixth the gravity we feel on Earth's surface, to keep things on the bottom. Still, weight is not as important; the partner beneath is not as likely to be pressed into breathlessness. One can rise higher in the air for a given effort and take longer falling so that there is more time for fun and games in midair than there could possibly be on Earth.

I suspect, though, that once zero gravity is experienced, mere low gravity will seem tame. Perhaps the time may come when large spaceships will be kept in orbit as sexual vacationlands.

Nor need we feel that sex in space is just a matter of an uninhibited orgy catering to man's animal instincts. Once colonies are established off the Earth, sex will still be needed for propagation, but that will not be all. It may come to much more than that.

Consider that we don't know, as yet, what the effect of prolonged exposure to low gravity is on the human body. Men have survived a couple of months—but what about exposure over a period of years or lifetimes?

Even in mere months of spaceflight, muscles and bones have involved comparatively cramped capsules in which movement was limited. Was it the zero gravity that produced undesirable effects on bone and muscle, or was it the inactivity?

Suppose it turns out (and it might well do so) that the human body can survive under low gravity over prolonged periods—but only if there are regular sessions of strenuous exercise to keep up the muscle tone and place the bones under necessary tension.

How can men and women be persuaded not to skimp on the necessary exercise? Is there any way that would be better than to choose some exercise that they would enjoy and *want* to do?

Why not, then, give them the opportunity of devoting an hour or two each day to sex—and the more prolonged, various, uninhibited, and playful, the better? It would be necessary therapy, but, of course, there would be no need to stress that.

Let's just consider it fun.

Chapter 33 • AFTERWORD

My writing (unlike myself in person) is so conventionally proper, and my fiction is so free of sexual motifs even in this day of X-rated everything, that I think I was invited to do the preceding essay partly because of the shock value of having my name on it.

And, indeed, after it appeared, a surprising number of my friends, with more or less salacious grins, saw fit to tease me about it.

But then came a time when the subject arose during a visit to a public school in Brooklyn. (It was in my old neighborhood, and I couldn't resist an invitation to talk to the youngsters there.) One young man, whom I judged to be about thirteen, said he had read the article.

"It's about the only thing of yours I've read that I didn't like," he said.

Alas, I thought, I am disillusioning my young readers, who have had such faith in my purity. "Why didn't you like it?" I asked, despondently.

"Not dirty enough," he said.

I rest my case.

34 · *Communication by Molecule*

The general trend in electronics is toward miniaturization. Once it required a large vacuum tube to control electron flow, but such a tube gave way to the much smaller transistor. Since the transistor was first announced to the world in late 1947, it has grown steadily smaller and more reliable, and circuit elements have been introduced by the thousand into tiny chips that are barely visible.

Are we approaching a limit? Not yet. There is a computerlike device of proven practicality that handles information by means of units far smaller than anything man's technology has been able to handle. Imagine a hundred billion individual components, each so small that the whole device weighs no more than 3 pounds and is no more than 9 inches along its longest diameter.

We call the device the human brain.

The individual components of the human brain are the various cells that make it up; but the cell, tiny though it is, is far more than a simple on-off switch. Each cell is, rather, an intricate and well-organized complex of molecules, and it is these molecules that, in a way, are the units involved.

The key molecules present in all cells are those of proteins and nucleic acids. These molecules are large ones, made up of anywhere from hundreds to hundreds of thousands of atoms; and it is these molecules that are capable of storing and transmitting information in a sufficiently controlled manner to support all the manifestations of life. These molecules (large though they are as molecules) are too small to be seen by even the best optical microscope, and that is surely small enough to satisfy any miniaturist of today.

How do such molecules work?

To begin with, all the characteristics of any species of organism, and of any individual within that species, are controlled by the total interrelationship of the chemical reactions going on within its cells.

The individual chemical reactions within the complex are controlled by certain protein molecules called "enzymes." Each enzyme can cause particular chemical reactions to proceed very quickly—reactions which, in the absence of those enzymes, would proceed much more slowly.

By controlling the nature of the enzyme mix, then, the nature of the over-all chemical reaction pattern would be fixed. Within a particular fertilized egg cell, for instance, there would be a particular enzyme mix that would bring about a set of chemical reactions that would, in its turn, bring about processes that would lead from the egg cell to a codfish, or lobster, or ostrich, or giraffe, or man, or any other of the more than million living species of organisms. And this without a mistake; no codfish ever laid an egg that developed into a giraffe.

But what decides the nature of an enzyme? Each enzyme molecule is built up of a chain of twenty different varieties of "amino acids." These can be put together in a vast number of different orders. A hundred amino acids, including five of each variety, can be put together in 40,000,000,000,000,000,000,000,000,000, 000,000,000,000,000,000,000,000,000,000,000,000,000,000,000,000,000, 000,000,000,000,000,000,000,000 different ways. The substitution of an amino acid of one kind for that of another can change the nature of an entire long chain, even when that is the only substitution. The mere reversal of two amino acids, introducing a slight change in the order, can change the nature of the chain and alter its chemical properties.

Something in the cell, then, must be capable of overseeing the formation of enzyme molecules of fixed amino acid orders. If fertilized egg cell after fertilized egg cell is going to develop into a man and not a giraffe (or vice versa) and do so with sufficient precision to make one man very much like another and one giraffe very much like another, the control of the formation of enzymes must be very tight and very exact.

The information device that oversees enzyme formation is contained in the "chromosomes." These chromosomes, arranged in pairs within the central nuclei of cells, are carefully duplicated in the process of cell division so that every new cell has a set of chromosomes for itself. The fertilized egg cell has one chromosome of each pair from the mother and one from the father.

Each chromosome is made up of a long line of molecules of a type known as "deoxyribonucleic acid," usually abbreviated as "DNA." The DNA molecules are made up, in turn, of long chains of smaller units called "nucleotides." These come in four different varieties, with chemical names usually abbreviated as A, C, G, and T.

The nucleotides can be arranged in any order and in any proportions. To have one nucleotide replaced by another, or to alter the order even slightly, is to change the characteristics of the DNA molecule. There is as enormous a range of possible orders of nucleotides as of amino acids.

Nevertheless, each species has its own characteristic supply of DNA molecules of a given type, and each individual within the species has a set in which the order is not quite like those of other individuals. It is this that makes you and me both human beings and yet each distinct.

How are nucleotide arrangements kept fixed? It turns out that each DNA molecule is composed of two twined strands that fit each other as a key fits a lock. At the time of cell division, the strands untwine and each brings about the formation of a new companion that fits itself. The lock, so to speak, brings about the formation of a new key; the key, a new lock. Each daughter cell can then have its own set.

Occasionally, the new formations aren't perfect, and slight changes in nucleotide composition or arrangement take place. As a result of such "mutations," new cells form that don't have quite the characteristics of their progenitors, or young organisms that don't have quite the characteristics present in one or the other of their parents. The changes produced by random mutations help bring about evolutionary development.

Given DNA molecules characteristic of an organism, how is enzyme structure controlled? If one goes along the DNA molecules and groups the nucleotides three at a time, one finds sixty-four different possible combinations (ATA, GCT, TCA, and so on). Each of the sixty-four triples is equivalent to one or another of the twenty different amino acids. The information from the DNA molecules (that is, the order of the triples) is transferred to the site of enzyme manufacture by intermediary molecules called "messenger RNA" and "transfer RNA." The enzymes that are then formed reflect, in their structure, the makeup of one section or another of the DNA molecules in the chromosomes.

If technology is ever to duplicate the technique of storing and transmitting information by use of molecules, the fine details of the process must be understood. All chemical reactions involve changes in electrons, and it would be necessary, therefore, to understand the fine electron changes in the workings of DNA.

At the Bell Telephone Laboratories, one attack is to expose DNA and individual nucleotides to ultraviolet radiation. Electrons will absorb the energy of the ultraviolet light and enter what are called "excited states" in which they contain more energy than they normally do. They will not retain that energy long but will give it up as light of wavelengths different from that of the ultraviolet they had originally absorbed. This re-emission of light is called "fluorescence."

From the detailed nature of the fluorescence something can be deduced concerning the behavior of the electrons. Distinct differences in fluorescence characteristics are noted when nucleotides are lined up in different arrangements, for instance.

Continuing studies of this sort can yield their most direct dividends in showing how ultraviolet light and other energetic radiation alter the chemical nature of DNA molecules, producing mutations, cancer, and even death. To understand the mechanism is to increase the chance of developing techniques for protection against such changes.

In addition, and more fundamentally, such studies might bring enlightenment as to the details of how molecules such as DNA control electron behavior.

Of course, the DNA molecule is enormously complex, and to study its intimate electronic control is a vast undertaking. If simpler molecules also display such control, it might be useful to investigate them. The knowledge gained in that way can serve, perhaps, as a stepping-stone toward an understanding of the more complex case.

Consider hemoglobin, for instance. It is a protein molecule of average size, less than a tenth the size of a DNA molecule. Hemoglobin is found in the red blood corpuscle, and its functions are considerably less complex than those of DNA. Its most important action is to pick up oxygen molecules at the lung and to then give up those molecules to the various cells.

About 96 per cent of the hemoglobin molecule is made up of four chains of amino acids. The rest of it consists of four comparatively small structures called "heme." Each heme is a roughly square arrangement of atoms (mostly carbons and hydrogens, with a few oxygens and nitrogens) at the center of which is a single iron atom. It is the iron atom, specifically, that picks up the oxygen molecules.

Since each hemoglobin molecule contains four heme structures and therefore four iron atoms, each hemoglobin molecule can pick up four oxygen molecules.

Ordinarily one would expect that when a hemoglobin molecule picks up one oxygen, it would become more difficult for it to pick up a second, then still more difficult to pick up a third, and most difficult to pick up a fourth. That is what we would expect, judging from most chemical reactions.

Yet this is not the case with hemoglobin. When a hemoglobin molecule picks up an oxygen molecule, it can then pick up another oxygen molecule with *greater* readiness. It is as though hemoglobin grows hungrier the more it feeds, so that it is much more efficient as an oxygen collector than it would otherwise be.

But why does this happen? Apparently, the act of picking up

an oxygen molecule changes the chemical nature of the hemoglobin in a subtle manner, so that the remaining heme groups can more easily combine with oxygen.

The four heme structures on the hemoglobin molecules are connected by chains of amino acids stretching from one to another. It would seem that electron shifts induced by the addition of an oxygen molecule to one heme can propagate themselves all the way across the amino acid chain to another heme. There another electron shift can be induced that makes the as yet unoccupied heme readier to accept oxygen.

At Bell Telephone Laboratories, the nature of these changes is being studied by a technique called "nuclear magnetic resonance," usually abbreviated "NMR." In this technique, an atomic nucleus, particularly one of hydrogen, is kept in a strong magnetic field and exposed to radio waves. Such nuclei will absorb radio waves, and the particular wavelength that is most readily absorbed depends on the exact distribution of electrons around that nucleus.

By studying the NMR of hemoglobin before and after one or more oxygen molecules have been picked up, something about the electron shifts that take place is being determined.

The delicacy of the control is indicated by the fact that there are abnormal hemoglobins that do not behave as efficiently as normal hemoglobin does when it comes to picking up oxygen. Yet such abnormal hemoglobins may differ from the normal variety by a single amino acid out of 141 in each of two of the amino acid chains in the molecule.

If by means of such research studies as are now proceeding at Bell Telephone Laboratories, we should master the principles of molecular communication, we will become ready to take an enormous additional step downward in the scale of miniaturization.

Computers with molecular memories may be no larger than those now in operation, and yet have memory banks large enough to contain the accumulated knowledge of the human race. We could then easily envisage a central computer serving the nation, or even the world, and acting as the general reference library for the population.

Government bureaus, industrial firms, educational or research institutions, and even individuals might have access to such a computer and might be able to ask at any time for those nuggets of knowledge not to be found in easily accessible ordinary references.

The vastly increased ease and thoroughness of information flow along the arteries and veins of human society might give a further impetus to intellectual and scientific advance—in the same way that printing once served the same purpose.

A computer of moderate size might serve to encode all human beings on Earth. They might be listed by all the various categories: age, sex, height, weight, nationality, marital status, education, profession, hobbies, medical data, legal data, everything—and with everything kept continually up to date. Each individual would then be truly an individual, occupying a definite niche in a computer memory system, to be called up and considered at will.

To those of us brought up with the ideals and habits of a simpler society, a total computerization of mankind may seem repugnant, and yet it may be that only so can a multibillion-person technological world be run efficiently.

And if computers can be built so compactly, and if information can be handled so quickly and in such quantity, there will be room to imagine computers more complex, by far, than anything that exists now, yet without expecting them to be of exorbitant size. A computer might be designed that would not only store data and produce it at will, but that would, on request, sift and correlate such data and come to conclusions on the order of complexity expected of the human brain. It could, in short, be made to reason.

If a true scheme of molecular information can be put into use by mankind, what is to prevent a computer from being made as complex and versatile as the human brain, and perhaps no larger? Perhaps, by thoughtful designing, by the proper choice of molecules (including those not available in the brains of living organisms), and by the use of solid-state adjuncts, a brainlike computer could be built that would be stabler than the human brain —one that could resist higher temperatures and higher radiation levels, one that would be immune to fatigue or to hallucination, and so on.

With such compact computer brains, we can easily visualize that science-fictional dream, the intelligent robot.

Nor is there any reason to suppose that we must stop at a level of complexity and versatility equal to that of the human brain. Having reached that level, we can go beyond it, too, and imagine computers (or robots) of more-than-human intelligence.

But will we then be replaced? Will Homo sapiens give way to Robot sapiens, as once the reptiles gave way to mammals?

And if so? Is the endless climb of life from the primeval slime arbitrarily to end at our level and no higher? Is change to become changelessness to suit our self-love?

Might we not, instead, take the attitude that evolution progressed by slow, random change until there was developed a structure complex enough to add purpose to evolution? The human brain is what adds the purpose, and the new and superior computer intelligence is then the next step in a new kind of evolution.

Why regret it? Is the history of Homo sapiens such that he doesn't deserve to be replaced?

And what nobler and better thing can we leave as our memorial than that we passed on the torch to a greater than ourselves, which we ourselves had created?

35 · The Stages of Fusion

Energy shortage? Call it oil shortage. Oil is so simple a source that it has kept us from working hard enough at alternates—and the best alternate is "nuclear fusion," which we're at the point of making possible, and which will last as long as man does.

In fact, we have been using it all along, we and every other form of life, since it is nuclear fusion in the interior of the Sun that produces the light and heat that bathes the Earth and makes life possible.

Deep in the Sun, hydrogen, the smallest of all atoms, is fused to helium, the second smallest. Deep in the Sun, the temperature is in the millions of degrees, and at that temperature hydrogen atoms will collide and fuse, provided the hydrogen can be kept in place.

At superhigh temperature, under earthly conditions, hydrogen would expand into the thinnest of thin gases. The atoms making up that gas wouldn't collide then because they wouldn't find each other. At the Sun's center, however, the Sun's own gravitational field, the weight of all those layers of the Sun's enormous bulk, keeps the central hydrogen in place, squeezing it so tightly together that the hydrogen is denser and its atoms far closer together than ever happens on Earth.

Scientists can't duplicate the Sun's gravitational field on Earth. Instead, they make use of a magnetic field. The magnetic field squeezes samples of hydrogen together (not nearly as tightly as at the Sun's center) and keeps them together for a little time. While the hydrogen is held in place, it is heated to temperatures of millions of degrees.

To make fusion work on Earth in this way, the hydrogen must be squeezed together tightly enough, kept that way long enough, and heated high enough, all at the same time, to get the process started. For twenty-five years, Soviet and American scientists have tried to do this and have been inching closer and closer, without yet quite making it.

And after all this time, it looks as though physicists have been following the wrong road and that there is another, better way. Instead of holding the hydrogen together and heating it slowly, it may be better to let the hydrogen go and heat it *very quickly*. If the hydrogen is heated quickly enough, its atoms remain together simply because there isn't enough time for them to fly apart.

That very device worked as long ago as 1952. Suppose the heat is delivered by an atomic bomb. If hydrogen, in some form, is packed about it, that hydrogen is heated enormously, and before it can expand and get away, it undergoes fusion and produces a much greater explosion than the original bomb could have. The result is a "hydrogen bomb."

Of course, an atomic bomb must be a certain size or it won't work at all. Even at minimum size, the result is a devastating explosion, so it won't do to have atomic-bomb-triggered hydrogen fusion—not for peacetime uses, anyway.

But in 1960, the laser was invented. This is a device for converting energy into a beam of light that can be concentrated onto a pinhead or less. The total energy of the beam of light might not be very large, but if all of it is concentrated onto a tiny spot, that tiny spot receives energy at an enormous rate.

Imagine a small bubble of thin glass containing hydrogen under pressure, with a laser concentrating its light upon it. The bubble would start to crumple and the gas would push forcefully away in the opposite direction. But suppose you have several lasers all striking the bubble from different directions so that the pushes balance. What then?

Now the bubble stays in place and the gas starts to expand. It might take only a millionth of a second for it to expand, but in that millionth of a second, the temperature rises to many millions of degrees.

What is needed now are lasers that are large enough and powerful enough to raise the temperature of the hydrogen high enough and fast enough to start fusion going. That still can't be done right now; lasers big enough haven't yet been built—but we can make the task easier by not using ordinary hydrogen.

There are two rare forms of hydrogen called deuterium and tritium. Deuterium fuses more easily than ordinary hydrogen does, and tritium fuses more easily still. Tritium is radioactive, however, and exists on Earth only in traces. It can be manufactured by bombarding a light metal called lithium with subatomic particles called neutrons. That, of course, is an expensive procedure.

Deuterium is stable, but rare; only a single hydrogen atom out of 7,000 is deuterium. Even so, there is enough deuterium in a gallon of water (if the deuterium in it is isolated and made to undergo fusion) to yield as much energy as the burning of 300 gallons of gasoline—and there is something like 1,000 billion billion gallons of water in the world, just there for the taking.

What is being done is to use a mixture of deuterium and tritium. This will fuse more easily than deuterium alone, and it is less expensive than tritium alone.

If the hydrogen under pressure is a mixture of deuterium and tritium, the laser beams we have now are getting close to the job. In 1968, Soviet physicists first detected individual fusion reactions in such systems. Individual atoms were sticking together—but not quickly enough to set all the hydrogen to fusing.

Plans are under way to build bigger and better lasers, therefore, and to design them so that they can deliver energy more rapidly and with even finer pinpoint concentration.

It should work. The problem has been licked scientifically, and what is left is engineering. We know exactly what we must do; the problem is to work out the actual machinery with which to do it.

The initial investment is huge. Enormous laser beams must be used, consuming vast amounts of energy. It also takes energy to isolate deuterium from water, to manufacture tritium from lithium and neutrons, and to freeze the deuterium/tritium mixture to unimaginably intense cold.

Once fusion is ignited, however, bubble after bubble can be dropped into place, laser flash after laser flash can fuse them. The floods of energy developed will then serve to power the laser beams and all the other necessary machinery, with plenty left over for the outside world.

Once a mixture of deuterium and tritium fuses, it forms neutrons plus helium. Helium is the safest substance there is; no worries there.

The neutrons, however, can be dangerous, and most of the energy produced in deuterium/tritium fusion is in the form of very energetic neutrons, which will move outward from the fusing pellets in all directions. What is planned, then, is to surround the reaction chamber with a shell of lithium.

The neutrons will react with the lithium to produce tritium, which can be isolated and fed back into the fusion reaction. In this way, the dangerous neutrons are consumed and the reactor itself makes the tritium it must use.

The lithium is heated, and melts in the process of absorbing and reacting with the neutrons. The liquid lithium is kept from becoming too hot by circulation past water reservoirs. The water is itself heated in the process and is converted to steam, which can turn a turbine and produce electricity.

Fusion energy, produced this way, is just about endless. Deuterium and lithium, the only substances permanently used up in the process, exist in sufficient supply on Earth to last mankind for millions of years.

What's more, fusion energy would be much safer than the nuclear fission reactors that exist today. Fission produces large quantities of radioactive products that are extremely dangerous and must be disposed of very carefully. Fusion produces tritium and neutrons, both of which are used up again in the process. There might be some leakages in the fusion process, but the dangers of that are inconsiderable compared to those of nuclear fission.

Nor could a fusion reactor possibly get out of hand. In fission, a sizable quantity of fissioning matter must be used or the thing just won't work. Precautions must therefore be taken against a catastrophic runaway reaction.

Deuterium and tritium will fuse, however, in microscopic quantities at a time. If anything at all goes wrong, the influx of bubbles stops and so does energy production. It is an automatic and certain fail-safe system.

Of course, when we say the scientific problem is solved and all that is left is to work out the engineering, we can't dismiss that too lightly. The engineering difficulties are formidable, and even if the first controlled fusion ignition takes place in the laboratory tomorrow, it would probably take at least thirty years before actual large-scale fusion power stations are built and set in operation. We are talking about a twenty-first-century process, therefore.

By the time we can first reasonably expect a nuclear fusion power station to be in operation, the world's population is going to be about 7 billion at least, which is nearly twice the present figure, and the world's problems will have multiplied greatly.

Fusion power is not going to help mankind, if, before that power can really take hold, civilization breaks down under the stress of internal problems. And even if fusion power is in the field supplying all those billions in time, it will eventually fail if those billions continue to multiply.

Fusion power holds the promise of a new golden age for mankind *if*, and *only if*, population is brought under control. If it is . . .

We can visualize a number of large fusion-power stations dotting the shorelines of the continents, serving as huge electric-power generators. Electricity will be cheap and plentiful again.

Nor need fusion power mean merely electricity. We can use it to power the combination of water and carbon dioxide into hydrocarbons and oxygen. These hydrocarbons would be equivalent to the oil and natural gas we now use. They would prove *better* than the natural product we now have. They would be pure hydrocarbon to begin with, so there would be no pollution on burning. They would merely burn back into the water and carbon dioxide from which they were formed, and in the process would use up only the quantity of oxygen that had been released when they were made.

For that matter, we might even learn to manufacture edible material out of water and carbon dioxide, using fusion energy. We could power our own bodies as well as our machinery.

We can also look beyond the deuterium/tritium first-generation fusion power stations. Scientists and engineers are bound to improve the process and develop more sophisticated fusion devices.

Already, for instance, there is talk of using boron and ordinary hydrogen as fusion materials. If they fused, all that would be formed would be helium and heat and nothing more. The absolutely harmless helium would be released, and the heat would be used to turn water into steam and run the turbines.

Of course, boron and hydrogen are much harder to set to fusing than deuterium and tritium are. Perhaps, at first, a small quantity of deuterium and tritium will be used just to start things going—like a match to a pile of kindling.

Once it becomes unnecessary to handle tritium and neutrons in any but the tiniest traces, there will be no need for many of the precautions and difficulties that would plague the first-generation power stations. Second-generation power stations would be smaller, simpler, and far less expensive.

Another step forward would come when it was no longer necessary to use gaseous hydrogen in any form, since that would always require the expense of keeping it under pressure. Eventually, compounds that are solid at ordinary temperature would be used, solids made up of atoms that would fuse with each other when heated. The compound lithium hydride is something that might be used.

Once such a third-generation fusion power station is built, the freezing procedures would be eliminated and other steps toward the inexpensive would be taken.

How small can you make a third-generation fusion power reactor?

If helium and heat are the only products, so that no complicated precautions against the danger of radioactivity need be taken; if solid substances are used; if lasers are made smaller and smaller and the solids themselves are used in smaller and smaller fragments . . .

Well, we might begin to think of a "microfusion reactor."

I can imagine one no bigger, perhaps, than an automobile engine. It would have storage battery auxiliary, because something would be needed for the initial activation of the tiny lasers, which would themselves be no bigger than spark plugs, perhaps.

The lasers, set into action when the car starts would ignite the fusing fuel, one microscopic fragment at a time. The helium would be released in microscopic quantities so that no formal exhaust system would be required. The energy could be used to run a generator to recharge the battery and power the car electrically.

Such a car would be silent and nonpolluting, and one supply of fuel should last for its lifetime.

Similar engines could run trucks, buses, tractors, and ships.

Jet planes and rocket ships could also run without ordinary fuel. Right now, in such action-and-reaction vessels, fuel burns and turns into hot exhaust gas. The exhaust gas escapes through a narrow opening backward, and the jet plane or rocketship moves forward. Given a microfusion reactor, however, ordinary water could be turned into superhot steam, which would supply the exhaust.

Is there any catch at all?

Yes, there is. Fusion produces heat that would not otherwise exist on Earth ("thermal pollution"), just as burning coal and oil do. This additional heat, over and above that which arrives from the Sun, cannot be radiated away by the Earth without a slight rise in the planet's over-all temperature.

If we use fusion energy in amounts that produce, say, a thousand times as much heat as today's burning coal and oil, that will only raise the Earth's over-all average temperature by a degree or two, but that will be enough. It will be enough to melt the polar icecaps over a relatively small stretch of decades and drown the heavily populated coastal areas of the continents under two hundred feet of ocean water.

We must therefore use moderation, and yet . . .

Even with less than a thousand times the present energy output, we can surely still have a heaven on Earth—*provided that we control population.*

B · The Whole of Society

Chapter 36 • FOREWORD

My views of the future aren't always consistent, as the following group of essays probably shows. After all, I don't *really* know what's going to happen. I can only play what seem to me to be the probabilities, and there are times when I am more pessimistic (or optimistic) than at other times.

One of my pet optimistic visions concerns a computerized world but, alas, I can never seem to convince those of my friends who, like me, are political and social liberals, that I am correct in this. They tend to fear computerization.

Consequently, when an organization of bankers asked me to do an article on the future of banking and urged me to use my imagination, I produced the following essay, which is presented, very thinly, in the guise of fiction. I thought, with relief, that dealing with bankers, undoubtedly conservatives, I could computerize to my heart's content.

Not a chance! When the article was finished and sent in, they blanched to a man and sent it back.

So here it is, and I leave it to you. Am I a minority of one in this respect?

36 · Sis

I don't think any outsider had landed on the Planetary Computer Bank in fifty years. My guide told me so, at any rate.

He was a quiet man with a low, distinct voice, an expressionless middle-aged face, lanky, thin-boned. He was dressed in a one-piece fiber-mold. He did not offer me his name; certainly not his nominal symbol. He was only my guide.

He said, "By now the world has grown used to the Planetary Computer Bank, and forgotten its debt, perhaps. The long period of peace and security under which mankind has lived for so long has given rise to the luxury of suspicion. That is why you are here. You may report freely what you see and hear."

We were moving along slowly in a four-wheeled magnetocart that slid along the inner surface of the world on magnetically held frictionless runners.

And the bank *was* a world, though a tiny one. It was a sphere, over 10 kilometers in diameter, circling Earth at a height of 4,000 kilometers above its surface. What "gravity" I felt was a gentle push outward, away from its center, thanks to the centrifugal force produced by the rapid spin imposed upon it.

My guide said, almost as though he had read my thought, "The push is significant only at the equatorial inner surface. There is where we are housed. We are all natives of the lunar colony, of course, so we are adapted to a lower gravity than you terrestrials."

My first question was thoroughly frivolous. I said, "On my way up here I heard the shuttle crew refer to the bank as Sis. Why Sis?"

My guide smiled, "You might call it an affectionate name. The computer is so large, intricate, and complex that some playfulness is necessary if it is to be dealt with. And there is a story—I don't know how true it is—that when the Planetary Computer Bank was established over a century ago, it was called 'Switzerland in Space,' and the initials of that are 'Sis,' you see."

"Switzerland?"

"A mountainous section of what is now the European Region. It served as an international banking center in the days when computers were nothing more than glorified abacuses."

He waved upward. "Come, let us move into the depths."

The magnetocart moved smoothly up a wall, which became floor to me after a few moments of nausea. My guide seemed unaffected. We moved past smooth walls of transplastic, which could (the information booklet told me) be made either transparent or opaque, in any section or combination of sections, by remote-controlled molecular adjustment.

I watched but saw no signs of transparency; much less any section opened. I said, "Is anything being repaired? I mean any part of—uh—Sis?"

The guide shook his head gravely. "It's been a long time since we've had to repair her. Sis repairs herself. She takes note of her own signals and can detect error or malfunction at a level far too subtle for us to notice. She then repairs herself with a speed far too great for us to attain. In a little while I will have portions of the panels made transparent for you so that you will see the molecular valves that are now replacing the microchips. Sis is doing the replacement herself, of course."

"And what do *you* do? I mean the human beings here on Sis?"

"Nothing," said my guide. "We are here for maintenance, which is never necessary; or for emergencies, which never take place. Someday mankind will be content to supply raw material and leave Sis to herself."

After about fifteen minutes of moderately rapid movement, he stopped the cart. He made an adjustment on a small device he was holding, and the transplastic on either side vanished into clear see-through. Within was a maze of components meshed into surrealist beauty.

"I can't describe that," said my guide, "because I don't know how it works. It is the oldest portion of Sis, but, of course, she has modified it considerably. The sections added since those days, under her guidance, are much more sophisticated, and we believe that Sis is phasing out this section altogether. She is still expanding, of course, and won't need this. If this portion *is* phased out, we expect Sis to maintain it as a system memorial."

He looked at me as though expecting me to laugh. I kept a straight face and said, "Why?"

He said, "It is this portion that first took over the banking function of the system. Every human being in the solar system was tabbed with a set of symbols representing his identification—birthplace, birthdate, parentage, dwelling places, physical appearance, education—everything! His or her assets were included and, from then on, every transaction in the system was recorded, whether the purchase of a loaf of bread or the appropriation of funds for the engineering of the Martian surface. The shifts in assets, translated into energy units, was recorded moment by moment. The current state of assets of any individual, group, or region was always available to the individual, group, or region concerned, or to any *legitimate* investigative inquiry. The tax level was always calculated, and asset shifts from individuals to groups to regions to the Central System Council were carried out moment by moment. Taxation became invisible."

I clear my throat gently. "For that very reason, some say, taxation can be excessive."

"For what purpose?" asked my guide. "No money can find itself illegally into the pockets of any official as long as Sis is in control."

I said, "There is a feeling that privacy no longer exists."

"To an extent that is true," said my guide. "In order to fulfill the banking functions, Sis has had to extend her factual input steadily. Every individual in the solar system is coded for all matters. You cannot change your dwelling place, your job, your sex partner, the state of your health, or anything else of significance without its being reported and recorded."

"But what do we gain?"

"Everything," said my guide. "To know and control intimately all financial transactions is to guide society. Done properly, as Sis does it, mankind reaches a level of security, prosperity, and peace. Given the numbers of man and the complexity of society, no other way of attaining such goals is possible."

"And cannot this control be misused? That is the growing fear of mankind."

"It cannot. I want you to understand that."

"May not someone with access to the computer gain information he can use as a weapon? If not to enrich himself, then merely to feed his power? If not an individual, then a government bureau, or the government itself?"

"I do not deny that this might constitute a desire," said my guide gravely, "but it cannot be done. Sis herself uses the data. No raw data need be supplied to anyone except under the most pressing and extraordinary circumstances. And if raw data must be supplied, the method of their use must be recorded. That use, whatever it is, will produce a calculable effect on the data collected from that point on. If that effect is not noted, then the data were used otherwise, illegally, and Sis will take action to nullify the illegality."

"Has that ever happened?"

"Not in over a quarter of a century. When one is quite certain that any illegality will be detected, there is no use in making the attempt."

I thought about it. "How about you and the others who tend Sis? Can't you reorganize its instructions?"

My guide laughed. "It is over a generation since any human being or combination of human beings has been able to modify Sis's program at will. Modifications may be offered. Sis herself then determines whether they are acceptable in the light of all the information she has."

"Might she not be wrong?"

"She hasn't been wrong so far. Consider the smoothly functioning society of man that now exists."

I said, "But with all her power, might not Sis take over?"

My guide leaned against the transparent wall of the great Planetary Computer Bank, stared thoughtfully at the endless maze of banks within, and said, "But she *has* taken over. She is not under human control. So what? She does what she is designed to do, and we benefit."

"But haven't we lost our freedom?"

"We have lost our freedom to destroy. And we could have that if we were to abandon Sis."

I said, frowning, "Couldn't we abandon Sis and take our chances —for the sake of freedom?"

"It wouldn't be much of a chance. We would wipe out nine tenths of humanity in the disorders that would accompany the effort. Sis has made that quite clear. She has been asked."

"She's no impartial adviser. She could be lying."

"Would you be willing to take the chance?"

"Couldn't Sis guide us in an intelligent return to a preindustrial society? One with less comfort and security, perhaps, but with more freedom?"

"We have asked. It would take five hundred years, and when an agricultural society is restored and Sis is finally dismantled, mankind will inevitably develop industrialization once again and return to this level. Sis assures us of that, and I suspect she feels strongly enough about it to use her financial power to frustrate any attempt we might make in that direction."

There seemed no way of shaking him. "What if someone destroys Sis?"

"Who would want to? Who would gain by world chaos and destruction?"

"Someone might do it perhaps thoughtlessly, or out of insane malice."

"How could he penetrate the protective guard set up by a humanity fearful of world chaos and destruction, a guard run with perfect efficiency by Sis herself?"

"What if the destruction is not human in origin? A meteor?"

"Impossible. There are no objects ever sighted between the orbits of Mars and Venus whose orbits have not been calculated in advance over centuries of time. Small objects can be pushed aside or destroyed. In the case of large objects, there will be plenty of time to push Sis aside—or prepare another."

"Then Sis—the Planetary Computer Bank—is all-knowing, all-powerful, all-controlling, all-good. She is . . ."

I did not finish the sentence.

My guide merely smiled and said, "Tell mankind not to worry."

I went back to Earth to carry the message.

37 · The Case Against Man

The first mistake is to think of mankind as a thing in itself. It isn't. It is part of an intricate web of life. And we can't think even of life as a thing in itself. It isn't. It is part of the intricate structure of a planet bathed by energy from the Sun.

The Earth, in the nearly 5 billion years since it assumed approximately its present form, has undergone a vast evolution. When it first came into being, it very likely lacked what we would today call an ocean and an atmosphere. These were formed by the gradual outward movement of material as the solid interior settled together.

Nor were ocean, atmosphere, and solid crust independent of each other after formation. There is interaction always: evaporation, condensation, solution, weathering. Far within the solid crust there are slow, continuing changes, too, of which hot springs, volcanoes, and earthquakes are the more noticeable manifestations here on the surface.

Between 2 billion and 3 billion years ago, portions of the surface water, bathed by the energetic radiation from the Sun, developed complicated compounds in organization sufficiently versatile to qualify as what we call "life." Life forms have become more complex and more various ever since.

But the life forms are as much part of the structure of the Earth as any inanimate portion is. It is all an inseparable part of a whole. If any animal is isolated totally from other forms of life, then death by starvation will surely follow. If isolated from water, death by dehydration will follow even faster. If isolated from air, whether free or dissolved in water, death by asphyxiation will follow still faster. If isolated from the Sun, animals will survive for a time, but plants would die, and if all plants died, all animals would starve.

It works in reverse, too, for the inanimate portion of Earth is shaped and molded by life. The nature of the atmosphere has been changed by plant activity (which adds to the air the free oxygen it could not otherwise retain). The soil is turned by earthworms, while enormous ocean reefs are formed by coral.

The entire planet, plus solar energy, is one enormous intricately interrelated system. The entire planet is a life form made up of nonliving portions and a large variety of living portions (as our own body is made up of nonliving crystals in bones and nonliving water in blood, as well as of a large variety of living portions).

In fact, we can pursue the analogy. A man is composed of 50 trillion cells of a variety of types, all interrelated and interdependent. Loss of some of those cells, such as those making up an entire leg, will seriously handicap all the rest of the organism; serious damage to a relatively few cells in an organ, such as the heart or kidneys, may end by killing all 50 trillion.

In the same way, on a planetary scale, the chopping down of an entire forest may not threaten Earth's life in general, but it will produce serious changes in the life forms of the region and even in the nature of the water runoff and, therefore, in the details of geological structure. A serious decline in the bee population will affect the numbers of those plants that depend on bees for fertilization, then the numbers of those animals that depend on those particular bee-fertilized plants, and so on.

Or consider cell growth. Cells in those organs that suffer constant wear and tear—as in the skin or in the intestinal lining—grow and multiply all life long. Other cells, not so exposed, as in nerve and muscle, do not multiply at all in the adult, under any circumstances. Still other organs, ordinarily quiescent, as liver and bone, stand ready to grow if that is necessary to replace damage. When the proper repairs are made, growth stops.

In a much looser and more flexible way, the same is true of the "planet organism" (which we study in the science called ecology). If cougars grow too numerous, the deer they live on are decimated, and some of the cougars die of starvation, so that their "proper number" is restored. If too many cougars die, then the deer multiply with particular rapidity, and cougars multiply quickly in turn, till the additional predators bring down the number of deer again. Barring interference from outside, the eaters and the eaten retain their proper numbers, and both are the better for it. (If the cougars are all killed off, deer would multiply to the point where they destroy the plants they live off, and more would then die of starvation than would have died of cougars.)

The neat economy of growth within an organism such as a human being is sometimes—for what reason, we know not—disrupted, and a group of cells begins growing without limit. This is the dread disease of cancer, and unless that growing group of cells is somehow stopped, the wild growth will throw all the body struture out of true and end by killing the organism itself.

In ecology, the same would happen if, for some reason, one particular type of organism began to multiply without limit, killing its competitors and increasing its own food supply at the expense of that of others. That, too, could end only in the destruction of the larger system—most or all of life and even of certain aspects of the inanimate environment.

And this is exactly what is happening at this moment. For thousands of years, the single species Homo sapiens, to which you and I have the dubious honor of belonging, has been increasing in numbers. In the past couple of centuries, the rate of increase has itself increased explosively.

At the time of Julius Caesar, when Earth's human population is estimated to have been 150 million, that population was increasing at a rate such that it would double in 1,000 years if that rate remained steady. Today, with Earth's population estimated at about 4,000 million (26 times what it was in Caesar's time), it is increasing at a rate which, if steady, will cause it to double in 35 years.

The present rate of increase of Earth's swarming human population qualifies Homo sapiens as an ecological cancer, which will destroy the ecology just as surely as any ordinary cancer would destroy an organism.

The cure? Just what it is for any cancer. The cancerous growth must somehow be stopped.

Of course, it will be. If we do nothing at all, the growth will stop, as a cancerous growth in a man will stop if nothing is done. The man dies and the cancer dies with him. And, analogously, the ecology will die and man will die with it.

How can the human population explosion be stopped? By raising the deathrate, or by lowering the birthrate. There are no other alternatives. The deathrate will rise spontaneously and finally catastrophically, if we do nothing—and that within a few decades. To make the birthrate fall, somehow (almost *any* how, in fact), is surely preferable, and that is therefore the first order of mankind's business today.

Failing this, mankind would stand at the bar of abstract justice (for there may be no posterity to judge) as the mass murderer of life generally, his own included, and mass disrupter of the intricate planetary development that made life in its present glory possible in the first place.

Am I too pessimistic? Can we allow the present rate of population increase to continue indefinitely, or at least for a good long time? Can we count on science to develop methods for cleaning up as we pollute, for replacing wasted resources with substitutes, for finding new food, new materials, more and better life for our waxing numbers?

Impossible! If the numbers continue to wax at the present rate.

Let us begin with a few estimates (admittedly not precise, but in the rough neighborhood of the truth).

The total mass of living objects on Earth is perhaps 20 trillion tons. There is usually a balance between eaters and eaten that is about 1 to 10 in favor of the eaten. There would therefore be about 10 times as much plant life (the eaten) as animal life (the eaters) on Earth. There is, in other words, just a little under 2 trillion tons of animal life on Earth.

But this is all the animal life that can exist, given the present quantity of plant life. If more animal life is somehow produced, it will strip down the plant life, reduce the food supply, and then enough animals will starve to restore the balance. If one species of animal life increases in mass, it can only be because other species correspondingly decrease. For every additional pound of human flesh on Earth, a pound of some other form of flesh must disappear.

The total mass of humanity now on Earth may be estimated at about 200 million tons, or one ten-thousandth the mass of all animal life. If mankind increases in numbers ten thousandfold, then Homo sapiens will be, perforce, the *only* animal species alive on Earth. It will be a world without elephants or lions, without cats or dogs, without fish or lobsters, without worms or bugs. What's more, to support the mass of human life, all the plant world must be put to service. Only plants edible to man must remain, and only those plants most concentratedly edible and with minimum waste.

At the present moment, the average density of population of the Earth's land surface is about 73 people per square mile. Increase that ten thousandfold and the average density will become 730,000 people per square mile, or more than seven times the density of the workday population of Manhattan. Even if we assume that mankind will somehow spread itself into vast cities floating on the ocean surface (or resting on the ocean floor), the average density of human life at the time when the last non-human animal must be killed would be 310,000 people per square mile over all the world, land and sea alike, or a little better than three times the density of modern Manhattan at noon.

We have the vision, then, of high-rise apartments, higher and more thickly spaced than in Manhattan at present, spreading all over the world, across all the mountains, across the Sahara Desert, across Antarctica, across all the oceans; all with their load of humanity and with no other form of animal life beside. And on the roof of all those buildings are the algae farms, with little plant cells exposed to the Sun so that they might grow rapidly and, without waste, form protein for all the mighty population of 35 trillion human beings.

Is that tolerable? Even if science produced all the energy and materials mankind could want, kept them all fed with algae, all educated, all amused—is the planetary high-rise tolerable?

And if it were, can we double the population further in 35 more years? And then double it again in another 35 years? Where will the food come from? What will persuade the algae to multiply faster than the light energy they absorb makes possible? What will speed up the Sun to add the energy to make it possible? And if vast supplies of fusion energy are added to supplement the Sun, how will we get rid of the equally vast supplies of heat that will be produced? And after the icecaps are melted and the oceans boiled into steam, what?

Can we bleed off the mass of humanity to other worlds? Right now, the number of human beings on Earth is increasing by 80 million per year, and each year that number goes up by 1 and a fraction per cent. Can we really suppose that we can send 80 million people per year to the Moon, Mars, and elsewhere, and engineer those worlds to support those people? And even so, merely remain in the same place ourselves?

No! Not the most optimistic visionary in the world could honestly convince himself that space travel is the solution to our population problem, if the present rate of increase is sustained.

But when will this planetary high-rise culture come about? How long will it take to increase Earth's population to that impossible point at the present doubling rate of once every 35 years? If it will take 1 million years or even 100,000, then, for goodness sake, let's not worry just yet.

Well, we don't have that kind of time. We will reach that dead end in no more than 460 years.

At the rate we are going, without birth control, then even if science serves us in an absolutely ideal way, we will reach the planetary high-rise with no animals but man, with no plants but algae, with no room for even one more person, by A.D. 2430.

And if science serves us in less than an ideal way (as it certainly will), the end will come sooner, much sooner, and mankind will start fading long, long before he is forced to construct that building that will cover all the Earth's surface.

So if birth control *must* come by A.D. 2430 at the very latest, even in an ideal world of advancing science, let it come *now*, in heaven's name, while there are still oak trees in the world and daisies and tigers and butterflies, and while there is still open land and space, and before the cancer called man proves fatal to life and the planet.

38 · The Son of Thetis

My father arrived in the United States in 1923, an immigrant from Eastern Europe. He had a high native intelligence, but no formal education of the type that would enable him to rise in the intellectual hierarchy of America.

He spent his life, therefore, as a candy-store keeper, and made it his ambition (as was common among immigrants) to see his sons get the education he lacked. The results were all he could have desired. His older son (myself) is a professor at a medical school and the author of many books. His younger son is city editor of a large newspaper.

His reaction to all this was one of unalloyed delight. When I pointed out to him, once, that had he had my education, he might easily have been me, he shrugged it off, and said, "There are two times when there is no possibility of jealousy: when a pupil surpasses his teacher and when a son surpasses his father."

With all possible respect to my father, I must say that I felt a certain anxious skepticism when he said this. It is all very well for my father, denied by circumstance the chance of making his mark in person, to be happy at making it vicariously. But what if he *had* had his chance, and had done quite well, and *then* saw himself surpassed by me?

Or suppose that I, myself, suddenly became aware that I was not, after all, entering literary history in my own right as Isaac Asimov—something that I have every reasonable expectation of doing. Suppose instead that I was right now coming to realize that I would, after all, enter it as a mere footnote—as the father of a much greater writer. As it happens, the situation does not arise, but I tell you frankly that if it had, I am not at all certain I would have felt my father's unselfish joy.

It is one thing to have something for nothing. It is quite another to have your own proud light go pale and sickly before the greater glory.

What would Philip of Macedon's reaction have been, I wonder, if after his quarter century of heroic striving, during which he raised his country from a backwoods nation of semibarbarians to the mastery of Greece, he had gained a sudden insight that he was destined to go down in history as "the father of Alexander the Great"? What about Frederick William I of Prussia, who in a quarter century of forceful rule built an awesome and frightening army out of a patchwork kingdom? What would have been his reaction if he had been made to understand that his place in the annals of man would be that of "the father of Frederick the Great"?

At that, they might have had some instinctive feeling of it, for each father hated his son, even to the point of threatening that son's life.

Hostility between royal father and heir-apparent son is commonplace, for there the conflict of present and future glory is all too obvious. Such hostility happens to be most traditional in the British royal family, dating back to the time when Henry II hated his sons (who were well worth his hatred) eight centuries ago.

The ancient Greeks, who thought of everything, took up the matter of the fear of the outshining glory of son or pupil in their myths and legends. Daedalus, the great craftsman and inventor of Greek tales, killed his nephew and pupil, Perdix, out of overwhelming jealousy, when that young man showed signs of becoming superior to his teacher.

More dramatic are the tales of the succession of supreme gods. The first ruler of the universe, in the Greek myths, was Ouranos. Him, his son, Cronos, castrated and replaced.

But once Cronos was seated on the throne, he was concerned lest he be served by his sons as he had served his own father. Therefore as his wife, Rhea, bore him sons, he swallowed each one in turn. When Zeus was born, however, Rhea fooled her husband by placing a stone in swaddling clothes, and that was swallowed instead.

Zeus was reared to manhood in secret and, in time, warred against his father, replacing him as lord of the universe.

There matters stood as far as the Greek myths were concerned, and yet Zeus was in danger, too. He and Poseidon (his brother, and god of the sea) both fell in love with the beautiful sea-nymph Thetis. They competed for the privilege of possessing her, until both hurriedly drew back on hearing that the Fates had decreed that Thetis would bear a son mightier than his father.

No god now dared marry the nymph, and Zeus compelled Thetis (quite against her will) to marry a mortal. The mortal was Peleus, and he was the father of Achilles, the great hero of the Trojan war, a son far mightier than his father.

In the light of this, it seems to me, it is not at all puzzling that people generally are afraid of robots generally. Why should not man fear the man-made man, the "son" of his hands, who may surpass him and prove mightier than his "father"?

Not so much man-made woman, you understand. In most early societies, women were considered inferior creatures who could not threaten man's priority. Pygmalion of Cyprus could fall in love with the statue Galatea, pray it alive, and marry her. Hephaistos, the Greek god of the forge, could have golden maidens minister to him in a counterfeit of life. Man-made *man*, however—the son, and not the daughter—was terrifying. Crete was guarded by a bronze giant, Talos, according to legend, who circled the island once a day and destroyed all outsiders who landed there. He had one weak spot, however, a stopper in the heel, which, if pulled out, would allow him to bleed to death. Jason and the Argonauts, on touching down at Crete on the way back from the adventure of the Golden Fleece, defeated Talos by pulling out that stopper.

To be sure, this is transparent symbolism. Crete, prior to 1400 B.C., was held inviolate by its bronze-armored warriors on board the ships of the first great navy of history, but the Greeks of the mainland finally defeated it.

However, there are all sorts of symbols that might be used to represent historical facts, and the Greeks chose to envision a mechanical man far more powerful than ordinary man, and one that could be defeated only with the greatest danger and difficulty.

The theme crops up over and over again throughout the legends of the ages. Man creates a mechanical device that in one way or another is intended to serve man within well-defined limits —and invariably the device oversteps the bounds, becomes too powerful, becomes dangerous, must be stopped, and scarcely can.

It is the case of the sorcerer's apprentice who brings the broom to life and then can't stop it. It is the case of the medieval rabbis who power golems of clay with the divine name, and then find that the power must be withdrawn, through difficulty and danger, before the manufactured man threatens the world.

In Christian times, a rationalization was advanced. A kind of life and intelligence could be created by man, but only God could create a soul. Any man-made man would be a soulless being, without the aspirations and moral understanding of a souled creature.

But this seems to me to be far too sophisticated to touch the point of the basic fear. Surely the mechanical man created to serve, but growing to surpass and endanger his creator, is the sublimated fear of the son, the beloved child who grows to surpass and endanger his father. Our fear of the robot is our fear of the son of Thetis destined to be stronger than his father.

Until the nineteenth century, that fear was only a whisper. Life could (in imagination) be imparted to inanimate objects only through divine intervention, entreated by prayer or enforced by magic. In 1798, however, the Italian anatomist Luigi Galvani discovered that the dead muscles of frogs could be made to contract by an electric shock. There seemed some connection between electricity and life, and the thought arose that life could be restored to dead flesh inside the laboratory and without the involvement of the unpredictable powers of the deities. The fear came closer and into sharper focus at once.

It was precisely Galvani's discovery that inspired Mary Wollstonecraft Shelley (second wife of the poet) to write her famous horror novel *Frankenstein*, published in 1818. A young anatomy student (in the novel) gathers together parts of freshly dead bodies and infuses them with electrical life. What he has created, however, is an eight-foot-tall monster of horrifying aspect.

Possessing intelligence and aware that he is forever cut off from human society, the monster turns upon the man whose interference with the course of nature has condemned him to solitary misery. One by one, the monster kills all of Frankenstein's family and friends, including his bride. Frankenstein himself dies of horror and remorse, and the monster disappears into the mysterious polar regions.

The book gave the language the phrase "Frankenstein's monster," now used for any creation that gets out of control, to the danger and horror of its creator. By its popularity, the novel sharpened the general suspicion that man-made man could only be evil—something that I, in my own writings, have referred to as "the Frankenstein complex."

Yet Frankenstein was written when science was in the flood tide of its vigorous, youthful optimism and when it seemed, to confident mankind, to be the ultimate answer to man's needs. It was not till World War I that science donned the mask of Strangelove horror. It was the warplane, and even more, poison gas, that showed mankind that the genius of the laboratory and of the inventor's workshop could be turned to death and destruction.

It is no accident that, soon after World War I, Frankenstein was out-Frankensteined. With inherently wicked man-made man constructed by a science that was itself capable of wickedness, it would not be only the creator that was threatened, but also all mankind.

In 1920 a play, *R.U.R.*, by the Czech playwright Karel Capek, was produced in Prague. In this play, man-made men were created as workers, to take over the muscle labor of the world, and to free men from Adam's curse at last. The inventor, Rossum, called his creation "worker." In the Czech language, the word is "robot," and this promptly entered the English language. The intials R.U.R. stand for "Rossum's Universal Robots."

It all works out ill. Men, without work, lose ambition and stop bearing children. The robots are used in war; they grow more complex and go mad; they rebel against mankind and destroy them. In the end only two robots are left. These exhibit human emotions, and it is through them that the world will be repeopled.

Mankind has been replaced by robots. Zeus has given birth to the mightier son of Thetis.

In the middle 1920s, the first science fiction magazine—the first periodical devoted entirely to the imaginative evocation of possible scientific futures—was published, and the era of modern science fiction began. With it there came an exploitation of the common motifs worked out earlier by such masters as Jules Verne and H. G. Wells.

Robots were not neglected. There were numerous tales of man-made man, but always, or almost always, the end was the same. The robot turned on its maker; the son grew dangerous to the father. Where this did not happen, it seemed almost as though the author were merely seeking a novel "twist," using the shock value of a kindly robot to produce a curiosity rather than to display the result of natural development.

That this wearisome parade of clanking monsters, forever parodying Shelley and Capek, came to an end was the result of certain stories written by myself.

When I began to write robot stories in 1939, I was nineteen years old. I did not feel the fright in the son-father relationship. Perhaps through the accident of the particular relationship of my father and myself, I was given no hint, ever, that there might be jealousy on the part of the father or danger on the part of the son. My father labored, in part, so that I might learn; and I learned, in part, so that my father might be gratified. The symbiosis was complete and beneficial, and I naturally saw a similar symbiosis in the relationship of man and robot.

Why should a robot hurt a man? It would be designed not to.

My first robot story appeared in the September 1940 issue of *Super Science Stories* and was entitled "Strange Playfellow." It dealt with a robot nursemaid named "Robbie." It was loved by the little girl she cared for but was distrusted by the little girl's mother.

At one point, when the mother expresses her concern, the little girl's father tries to argue her out of her fears. He says:

"Dear! A robot is infinitely more to be trusted than a human nursemaid. Robbie was constructed for only one purpose—to be the companion of a little child. His entire 'mentality' has been created for the purpose. He just can't help being faithful and loving and kind. He's a machine—*made so.*"

There you are. Already I had the dim notion that in the manufacture of a robot, a deliberate design of harmlessness would be built in.

This idea developed further. By the time I wrote my third robot story, "Liar!," I was ready to be more formal and precise about this matter of harmlessness. In "Liar!," published in the May 1941 issue of *Astounding Science Fiction*, one person says to another, "You know the fundamental law impressed upon the positronic brain of all robots, of course."

And the answer comes, "Certainly. On no conditions is a human being to be injured in any way, even when such injury is directly ordered by another human."

But then this cannot be all that must be impressed upon a robot's mind. By the time I wrote my fifth robot story, "Runaround" (which appeared in the March 1942 issue of *Astounding Science Fiction*), I had worked out my "Three Laws of Robotics." (Even the very word "robotics" is, as far as I know, my invention.) Here are the Three Laws in final form:

The Three Laws of Robotics
1. A robot may not injure a human being or, through inaction, allow a human being to come to harm.
2. A robot must obey the orders given it by human beings except where such orders would conflict with the First Law.

3. A robot must protect its own existence as long as such protection does not conflict with the First or Second Laws.

I am the only science fiction writer who actually quotes the Three Laws in fiction, but the popularity of my robot stories has been such that readers have come to take the Three Laws for granted. Other writers of robot stories tend to accept them and to write within the frame of the Three Laws, even though they do not state them explicitly. I am entirely happy over that.

To be sure, this is not an absolute requirement. In the motion picture *2001: A Space Odyssey,* and in the novel written from it by my good friend, Arthur C. Clarke, the complex computer Hal (a robot in the broad sense of the word) brings about the deaths of several human beings. This disturbed me quite a bit, and impressed me as a retrogressive step, but it doesn't seem to bother Arthur at all.

But what about computers? Even if we classify them as a kind of robot evolved to all-brain-no-body, and place them under the Three Laws, might they still not become uncomfortably complex and capable? Even if the son does not become dangerous to the father physically, might he not, with the best will in the world, become dangerous psychologically? Might he not force the father to admit inferiority? Might the father not be forced to hand over the universe to a kindly and regretful but inexorably demanding son?

There is a strong tendency on the part of those who secretly fear this to downgrade the possibility as, I suspect, a matter of self-protection.

The computer can*not* equal the human brain, is their feeling. The computer can*not* do any more than it is programmed to do. The computer can *never* exhibit the intuitive qualities of creativity and genius that the human brain can.

I wonder if there is not also a definite feeling, usually not expressed, out of a certain mid-twentieth-century embarrassment, that man has something called a soul that a computer cannot have; that a man is a product of the divine and a computer cannot be.

It's my opinion that none of these arguments is convincing.

The most advanced computer of today is an idiot child compared to the human brain, yes. But then, consider, that the human brain is the product of perhaps three billion years of organic evolution, while the electronic computer is, as such, only thirty years old. After all, is it too much to ask for just thirty years more?

What is to set the limit of further computer development? In theory, nothing. There is nothing magic about the creative abilities of the human brain, its intuitions, its genius.* It is made up of a finite number of cells of finite complexity, arranged in a pattern of finite complexity. When a computer is built of an equal number of equally complex cells in an equally complex arrangement, we will have something that can do just as much as a human brain can do to its uttermost genius.

To deny this is to maintain that there is something more in the human brain than the cells that compose it and the interrelationships among them.

And if human brain and man-made brain reach the same level of complexity, I feel it will be a lot easier to design a still more complicated man-made brain than to breed a still more complex human brain. So not only man-made man is possible, but man-made superman, too.

And how long will it take to reach the human brain level? A million years? A billion?

That, I suspect, is more consolation. Much less time, *much* less time may be required.

The key problem wll be this: to design a computer capable of formulating the design of another computer just slightly more complex than itself. Such a computer would naturally design another computer that was somewhat more capable than itself in designing another computer still more complex, which would be still more capable of designing still another computer even more complex, and so on.

We will be faced, then, with what mathematicians would call a diverging series.

Once the crucial moment arrives when a computer can design a computer greater than itself, computers will follow in rapid suc-

---

* I am always amused to hear some perfectly ordinary human being pontificate that a "computer can't compose a symphony," as though he himself could.

cession and rise out of sight. The son of Thetis will have been born.

And when will that crucial moment come? It might arrive long before the computer is as complex as the human brain. All we will require is a computer, however simple, to form another more complex than itself, however slightly. That will be the chain reaction that will produce the computer explosion—and the crucial moment may come next year, for all I know.

And what if it does? What if the computers show signs of getting away from us? Would we be face to face with a real Frankenstein's monster at last? Must we all struggle to destroy the thing before the divergence proceeds to the point where we are helpless before it?

Will the computers (oh, horrible thought!) *take over?*

What if they do? The history of life on Earth has been one long tale of "taking over." From era to era, different forms of life have proved dominant in one major environmental niche or the other. The placoderms "took over" from the trilobites, and the modern fish "took over" from the placoderms.

The reptiles "took over" from the amphibia, and the mammals "took over" from the reptiles.

Mankind looks upon the history of evolution and approves of all this "taking over," for it all leads up to the moment when man, proud and destructive man, has "taken over."

Are we to stop here? Is Ouranos to be replaced by Cronos, and Cronos by Zeus, and no more—thus far and no farther? Is Thetis to be disposed of rather than risk the chance of further replacement?

But why? What has changed? Evolution continues as before, though in a modified manner. Instead of species changing and growing better adapted to their environment through the blind action of mutation and the relentless winnowing of natural selection, we have reached the point where evolution can be guided and the successor can be deliberately designed.

And it might be good. The planet groans under its weight of 3.4 billion human beings, destined to be 7 billion by 2010. It is continually threatened by a nuclear holocaust and is inexorably being poisoned by the wastes and fumes of civilization. Surely it is time, and more than time, for mankind to be "taken over" from. If ever a species needed to be replaced for the good of the planet, we do.

There isn't much time left, in fact. If the son of Thetis doesn't come within a generation, or, at most, two, there may be nothing left worth "taking over."

39 · The Magic Society

We all have our dreams, and surely one of them is to live forever. Where is the man or woman, with a life that has the least bit of sweetness in it, who had not, on one occasion or another, consciously regretted the necessity of death? Who does not have the tendency to cling to life, even when that life is far from perfect?

Immortality is a magic word, and any society composed of immortals is a magic society. Or at least it might be. . . .

For instance, is immortality, all by itself, necessarily a blessing? The souls of men in hell are pictured as immortal, but such immortality is the ultimate curse.

We need not go to such extremities, however, to demonstrate that the fact of immortality is not, of and in itself, necessarily desirable. Consider one of the most famous legends of all time, that of the Wandering Jew. For his heartlessness on the occasion of the crucifixion, he was given what all would agree to be a horrible punishment. Yet that punishment consisted of nothing more than the grant of immortality.

There is no suggestion that the Wandering Jew was condemned to suffer any unusual pain or was deprived of any necessity of life. He was merely condemned to wander forever. Yet traveling can be fun, and immortality seems to be desirable. Why then should an eternity of traveling be looked upon as such a horrible punishment?

The answer lies in constraint. The Wandering Jew *must* wander. For him there can be no rest; the horizon always beckons.

If, then, we are to build our Magic Society, we must include immortality and we must exclude constraint.

In fact, immortality is itself the consequence of the absence of constraint. No one need necessarily grow old and die. For that matter, no one should necessarily be sick or feel pain. We want not only immortality but also eternal health and eternal youth.

Then, too, no one in the Magic Society ought to have to grub for a living. Who has not wished, at one time or another, to be boundlessly wealthy, as the easiest method of insuring that one's wishes will be gratified? To life and youth and health, let us add leisure—leisure to pursue one's interests and pleasures.

Can anything more be needed? Yes, a bit more.

We must assume that an immortal human being in our Magic Society still has the emotions of a human being; that he would have the capacity for loving his mate, cherishing his children, enjoying the company of his friends. If he alone were immortal, his life would consist of the ever-recurring loss of loved ones, the accumulation of aching memories.

The perfection of happiness cannot exist if confined to one person, or even to any small group of people. Unhappiness can invade from without unless there is no without. In short, the Magic Society must include perpetual life, health, youth, and leisure for all human beings. And since health includes mental and emotional adjustment, there would, ideally, be no malice, cruelty, or envy that would make the happiness of one depend on the unhappiness of another. The Magic Society would be good as well as happy.

The Magic Society, as I have described it, seems to be what our physicians, psychiatrists, scientists, and engineers are striving for —a world where old age is conquered, minds are adjusted, and all work is done by robots at the expense of limitless energy obtained from the copious nuclear stores.

But what I have described is also similar to the popular conception of heaven—a place where the souls of good men and women live forever in the absence of all that is evil, where all days pass in bliss, and where even the slightest discomfort or uneasiness is unknown.

If we are striving for heaven on Earth and attain it, is that all? Happy ending? Drop the curtain?

I wonder.

When I was quite young, I read a play concerning a man who died and came to consciousness to find himself surrounded by utter beauty and ultimate comfort, with the perfect servant at hand to wait upon him.

Complacently, our hero decided that such had been his good life that he deserved this reward and promptly began to enjoy all the delights that were made available to him.

Years passed while he enjoyed the company of the most fascinating men and the most beautiful women in history, savored the great works of art and literature, listened to the best music played by the most skillful musicians, amused himself by observing the passing parade on Earth itself, and indulged himself in his own creative efforts in one field after another. Nothing could be more desirable and wonderful.

Yet occasionally, he would grow pensive. He did once in a while miss the chance to do some physical work. There was an occasional feeling of monotony to bliss. Whenever he asked the chance to do a little ditch digging, however—just for the fun and novelty of it—his supernatural servant shuddered and said that such things could only be found "in the other place."

Time passed, and our hero grew to long more and more for work, even for the experience of weariness or pain, but this was always refused as belonging only "to the other place."

And when boredom grew to such a pitch that he could bear it no more, he cried out, "If I am condemned to eternal happiness and nothing more, then I would rather be in hell!"

And the supernatural servant smiled urbanely and said, "Why, sir, wherever do you think you are right now?"

In short, in considering the Magic Society, we have left out of account the most painful and incurable of all disease—boredom.

It is possible to argue that even the Judeo-Christian God of the Bible must suffer from the boredom that accompanies perfection.

"What," someone asked of St. Augustine in about A.D. 400, "was God doing before he created heaven and Earth?"

"Creating hell," roared back St. Augustine at once, "for people who ask questions such as that."

And yet this has always struck me as a legitimate question. God had already existed an eternity, and did He not find it irksome or boring to be alone and to do nothing?

That might seem to be a naïve question that would arise only out of the imperfection and incompleteness of the human mind. *We* would be bored if we were alone for an eternity, for our thoughts are limited and our capacity for creativity small. God, on the other hand, being infinitely complex could, one might suppose, contemplate Himself for an eternity without growing bored. The word itself would lack significance applied to Him.

Yet, if that were so, why did He create mankind? Could it have been for any other purpose than to give Himself something to do?

Consider the nature of His creation. Had He created mankind perfect, He would have gained nothing. He Himself was already perfect and complete, and to create a perfect mankind would have merely meant duplicating a part of Himself.

Instead, then, He created man as a highly *im*perfect being— one that was so imperfect that the Bible is little more than an account of God's failure to persuade man to become a little less imperfect.

No sooner are Adam and Eve created than they disobey the one small rule God ordered them to obey. Their oldest son becomes a murderer, and after ten generations mankind is so sunk in iniquity that God must flood the Earth and start over. But the generations after the Flood quickly corrupt themselves, and God cuts his losses by confining his main effort to a single group, the descendants of Abraham. Even these backslide continuously, despite the closest supervision, and must be punished from generation to generation in a variety of ways without its ever, apparently, doing them permanent good.

Nor, in view of the state of mankind today, can it be convincingly argued that the coming of Christ two thousand years ago made a significant dent in the imperfections of man.

To suppose that man's imperfections are the result of God's not being able to do any better does not jibe with the general belief among Jews and Christians that God is omnipotent and can do anything. One might prefer to suppose, instead, that God carefully designed this imperfection for His own purposes and that only by endlessly manipulating a world carefully designed to remain imperfect could He stave off boredom.

Let us therefore add one more feature to our Magic Society and remove the last of all constraints—the constraint to be perfectly happy. In the play I described, our hero *had* to be happy, he *could not* experience pain or sorrow, and of course he was *forced* to live on and on. It was that which built hell in heaven's despite. God, wiser than that, deliberately marred perfection by creating imperfection.

In the Magic Society toward which mankind is now tending, we must therefore have certain imperfections exist, and surely they will be ready to hand.

For instance, the possibility of death by accident always exists. While a man is constructed of flesh and blood, however perfectly adjusted that might be, he can always be killed by being run over by a steamroller.

It has been estimated, in fact, that if accident were the only cause of death, and if these accidents occurred at the rate they now occur, then the average life expectancy would be two thousand years. The bored of the Magic Society can, therefore, look forward to eventual release even if they make every effort to live.

But would they necessarily make every effort to live? Consider the problem of population stability. Clearly, if each long-lived couple continues having children at the rate people now have them, and if those children, once they reached maturity, had children of their own and so on, then the Earth's surface will be crowded past endurance before half the original long-lived individuals manage to escape from life.

The only tenable solution is to have the population hold at some not-to-be-exceeded level. If the average couple indeed lives for two thousand years, then they must have, on the average, no more than two children in the course of those two thousand years, which means one child per millennium. Let us put it another way. In the Greater Boston area, there would be no more than one thousand children less than a year old in any given year.

Children would be a rarity indeed in the Magic Society, and having a child of one's own an even greater rarity.

This is something to regret. It is not just that children are cuddly or sweet or that taking care of them fills a deeply felt human need. (I'm not very sentimental about children, myself.) It is rather a quality of mind that will disappear from humanity.

As a person lives and accumulates experience, he carves out a way of thought for himself and settles into comfortable mental ruts. This will be more than ever so for people who measure their ages not in decades but in centuries. It is the young people who approach the problems of man and the universe with fresh minds and unorthodox views, and it is they who are responsible for most of the vigor and verve of mankind. One could go through the history of science, for instance, and show how greatly youngsters in their twenties have contributed to scientific advance and how little oldsters in their forties have.

A world of cram-packed minds is a dull one to consider, and the absence of freshness would be an unimaginable tragedy.

Yet if we are to increase the number of the young, one cannot avoid decreasing the number of the old; and if accident does not decrease the latter quickly enough, the most civilized alternative is to encourage suicide.

This would undoubtedly not be as difficult or heartless as it might sound. Even now, though life is short and death comes soon enough without help, about one man in ten thousand commits suicide in the United States each year, and the rest experience the impulse now and then. With extended life spans, despondency over the accidental death of a friend or loved one could urge a man to suicide, as could (much more often) the sheer gathering boredom of it all.

I can visualize government-regulated, robot-staffed suicide stations where individuals can spend a few final weeks in comfort, entertaining friends and family, taking care of any final tasks that need doing, read that one last book they always meant to read, have a final opportunity to change their mind, and then be helped to drift softly into the final sleep.

Such sponsored suicide might not only solve the population problem but also change the very nature of the race. Suppose we consider the problem of work, or, rather, the lack of work.

What would it be like if the whole world were one huge leisure class? We might try to find an answer by considering the leisure classes we have now and have had in the past. Some members of the class engage their lives in one long bout of time wasting, in a thinly disguised wait for death. Sometimes the wait is not so thinly disguised either, as one sees when one considers the popularity of the danger sports among them: dueling in the past, racing cars in the present.

Others, however, make work for themselves, sometimes highly constructive work. They have been scholars, educators, scientists, statesmen, military leaders. There have been times when governments have been run by the leisure class almost exclusively, and there was a period early in the history of modern science when a scientific revolution was carried through almost entirely as a leisure class activity.

With the whole world a leisure class, no doubt the same split between the two possible responses will occur.

The overwhelming preoccupation of part of the population will be that of finding some way of "having fun," of passing the time pleasantly, enjoyably, excitingly, even dangerously. No doubt there will be kinds of "fun" in the future we don't have today—telepathic fun and games for all I know, space flights for amusement, many things.

Some, on the other hand, will react creatively. There will be a strong impulse on their part of "make work." The simplest method and the one most available might simply be to order the robots back and take over, at least to do so once in a while for the fun of it.

Unfortunately, I don't think that any really significant work could be done in this fashion. As technology advances, it becomes more and more intricate, and the random insertion of some unexpected variable can do damage on a larger and larger scale. I need only mention the great Northeastern electrical blackout of November 9, 1965.

After a few unfortunate experiences, society will strongly resist attempts to disconnect any important segment of the robotic network, and that would leave ordinary labor out of the question as a safety valve for boredom.

There will remain, of course, the world of scholarship—which can be expected to remain peculiarly human, however omnipresent and versatile robots become. Earth can be filled with gentlemen scientists, gentlemen antiquarians, gentlemen chess enthusiasts, gentlemen artists, gentlemen writers, and gentlemen collectors.

And yet I feel that while men remain essentially as they are today, the division into these two groups—one of which requires amusement from without, the other from within—can never be equal.

The outer-directed group, it seems to me, will be much in the majority, and they are bound to find that amusement from outside—whether sex, TV, or games now unimagined—is, essentially, a long-run failure. Nothing bores so surely, so easily, and so irreversibly as amusements plastered onto the surface of an impermeable mind.

The inner-directed men, on the other hand, can create a personal world of interest designed for themselves. They need not be great artists to do so. Their individual worlds may even seem inconsequential and dull to others, but what comes from within can brighten and vitalize an individual and make life not only bearable but also desirable.

Even amusements from within cannot make life tolerable for an eternity, but they can do better for longer than amusements from without can. At all age levels, the percentage of outer-directed men frequenting the suicide stations will be higher than the percentage of inner-directed men. There is thus bound to be a shift in the nature of the population in the direction of the inner-directed.

Furthermore, I imagine that the tendency will be progressive. Suicide will be increasingly popular among the outer-directed, to the point where the population can begin to decline.

I suspect that the decline will not bother people. There will be arguments in its favor. A leisure class (I can hear scholars argue) is best and most profitably sustained when it is a cultured minority. In the past such leisure classes, even the most admirable of them, have been built on the backs of slaves, as was the Athens of Pericles, or on the backs of ground-down peasants, as was the Paris of Voltaire.

In the Magic Society, the robots take the place of the slaves, and there is no need for a vast population of individuals—a cultureless leisured class—ill adapted for the Magic Society, who exist only as hangovers from the nonmagic societies of the past.

Why not, then, allow population to decline selectively and create a new Periclean Athens, with the cream of mankind at the apex of a pyramid of robots?

I once depicted such a society in a novel of mine called *The Naked Sun*. Each man, or woman, had his or her own vast estate in a world run by robots. I pictured this world as one in which it was the mark of a civilized existence to be able to travel a leisurely day without encountering any other human being. It was vulgar in the extreme ever to be in the position of physically seeing someone. All necessary contact was done by three-dimensional closed-circuit television, which was called "viewing" as distinct from "seeing."

Each individual in such a society could take for his province some specialized segment of scholarship and become pre-eminent in it during the course of a long, leisurely life. And yet murder entered this particular society, and I made it plain that there was within it the seeds of instability.

After all, life has been a to-the-death gamble all through history. No man has been secure from the sudden and unexpected onslaught of accident, disease, and malice (acts of God, we call them). In the Magic Society, however, disease and malice will not exist, and even accident will decline as the number of people dwindles and the number of robots increases.

In a more subtle sense, life is a to-the-death gamble, too. In the teeming population of today's society, made up of fiercely self-seeking individuals, a man has perhaps thirty mature years in which to impress himself on the world and to "succeed." Most or all of us take part in this thrilling game, whether we know it or not and whether we are consciously competitive or not.

I'll give you an example. A scientist friend of mine was telling me the other day that he had just successfully worked out a problem that had stymied research workers in his field for decades. As he grew more enthusiastic, his eyes glowed mildly; a soft chuckle made itself heard, and he said as he rubbed his hands, "My colleagues will *hate* me!"

This is the urge for "glory" in all its aspects.

And what will become of glory in the Magic Society? You have centuries to gain whatever you wish. In a dwindled population, you need have no competition in your chosen field of work. Since you have no significant chance of losing, there is no significant value to winning—and nothing can make up for that.

I have a feeling that even after the Magic Society has attained everything—long life, total leisure, a uniformly cultured population—men will continue to decline in numbers, for there will be nothing in the long run to make life worth the trouble. Homo sapiens will finally die the quiet death of peaceful old age, but there are, after all, worse things than that.

And thereafter, the Earth will continue to hum busily along— at the hands of self-repairing and self-perpetuating robots.

Perhaps the day will come when the robots come to write histories of their own. Perhaps they will forget (or will never have realized) that they are robots but will consider themselves men and will be so complex in structure that they might as well do so. They might construct legends of past Golden Ages in which demigods inhabited the Earth without sickness or pain or grief or labor, and where the only death was a quiet going off into sleep. (The Greek poet Hesiod made up exactly such a legend.)

The robots might long for the Golden Age and wonder how it had come to be lost. They might even be driven to invent explanations.

"There was this serpent . . ." they might begin.

Chapter 39 • AFTERWORD

The immediately preceding essay and the one immediately following deal with my attitudes toward immortality. In both essays, in different ways, I express my opposition.

Sometimes I wonder if I am being hypocritical. Would I, after all, refuse immortality if it were offered me? I guess not. Whatever my intellectual attitude toward immortality, I am not eager to die. I experience too many good things in life to *want* to die.

However, no one is offering me immortality, and at the present time my age is three fourths of what my father's was at the time of his death. With only a quarter of my life left (and the worst quarter), I find I am not unduly depressed at the inevitable end.

It is *not* a tragedy to die. It *is* a tragedy to have lived a useless life. I have done my best to avoid the latter.

40 · To Life
—But Not Forever

Who's for immortality? Who wants to live forever?

There's a way, some people think. First you die; but immediately after death, you get yourself frozen in liquid nitrogen as quickly as possible. Once you're colder than cold (liquid nitrogen will keep you unchanged at roughly 200° below zero, Centigrade, which is the equivalent of 320° below zero, Fahrenheit), you will deteriorate no further; you will be no further dead than you were when you died.

Dead enough, you say? But wait. There will come a time, a hundred years hence, five hundred years hence (who cares? your liquid-nitrogen-frozen body will keep indefinitely) when whatever it was that killed you will be curable. The doctors will bring you back to life and cure you. You can then go right on living.

What have you got to lose? You won't have to live tediously through the waiting. However long it may be, it will seem to you to have passed in a flash. You will close your eyes in death and open them in life in the space of a wink. In fact, the longer the wait, the more interesting the future into which you enter.

Of course, it will cost a little money. Everything does. It will cost $30 a year to belong to the society that arranges this; $8,500 (I am told) for the initial freezing, and $1,000 a year to maintain it. And maybe expenses will go up with time; everything does. But what of it? So you lose a little money. How does that compare with your life?

Any catch?

Well, let's see. Once you're dead, are you sure that your descendants will keep your freeze coffin going, if it takes a really long time for the doctors of the future to get around to you? Your sons and daughters might be filial enough to devote $1,000 a year on your frozen meat; maybe your grandchildren, too. However, humanity being what it is, how far can you trust your great-grandchildren and great-great-grandchildren?

That problem has a solution, however. You merely establish a trust fund, which will be designed to keep you frozen in perpetuity. That amount of money, of course, will not be available for passing on to your children, and the young skunks may be ungrateful, so don't put them in charge. Pick some impersonal bank.

Then the kids will in turn get their own bodies frozen, each with a trust fund, and more money will be removed permanently from the family fortune, though that's not *your* worry.

In fact, if this sort of thing gets popular (and why shouldn't it?) and if more and more people decide to freeze themselves—thousands, millions—it may be that the world will become full of banks upon banks of cylinders, all filled with liquid nitrogen, all outfitted with refrigerating devices keeping them forever cold and consuming large quantities of energy.

All of us will be watching our dead carefully, in the hope that our own descendants will watch us just as carefully. It will be a death-centered society, like the Egyptians, except that they only built pyramids, and used chemical mummification instead of liquid nitrogen mummification.

Of course, we won't be death-centered, will we? We'll be expecting the frozen corpses to be brought back to life, won't we? (Yes, but so, as a matter of fact, did the Egyptians.)

And that means the number of freeze cylinders won't increase forever. Someday doctors will start bringing the dead back to life, won't they?

If we wanted to be pessimistic, we could say that although some simple cells (even some simple human cells) can be frozen while in a state of vigor and then brought back to life, no one has succeeded in doing it to large organisms of anywhere near the complexity of man (especially of a man who is not in youthful vigor, but who has just died of a disease, or an accident, or old age, that has reduced his body to a state of serious disorganization). There is no real hope, in fact, that it can ever be done.

But that is pessimism, bad thinking. What we need to do is turn our thoughts upward. Someday, we must repeat to ourselves over and over, mankind will learn to restore the spark of life to a body, however decayed, however smashed, however disease-riddled.

Let's even guess when this might happen. Let's say A.D. 2500. That's a nice round year, not so close as to be ridiculous, not so far away as to be hopeless. It means that people frozen now will have to pass away their dreamless sleep for a little over five centuries. Not bad.

How many cylinders of frozen bodies will be around by 2500, do you suppose?

There may very likely be only a few. The process is expensive, and as the years go on, people may lose heart. Perhaps only a thousand will have been frozen and maintained and waiting. In that case, the whole movement will not have been important.

But suppose, on the other hand, the process grows popular, becomes a runaway fad, that methods for freezing wholesale and cheaply are evolved, that national trusts are set up for maintenance. Even so, let's not suppose *everybody* gets frozen. Let's suppose only one out of a hundred do.

If that is so, we must remember that right now about 70 million people die each year in the world, and that if this figure stays the same every year up to 2500, then by that year, there will be 370 million people soaking in liquid nitrogen and waiting for the blowing of the trumpet of some secular Gabriel. That's one tenth the present population of the world.

So 2500 comes and the doctors now have the ability to restore life to the 370 million waiting. The question is, will they?

We can see advantages in their doing so, of course. They could revive some individual from 1972 and another from 1998 and still another from 2045, and so on. Think of the historical value of probing their memories; the firsthand information they could get of past mores, customs; the reminiscences of famous men who influence history (if those men were frozen; or of reminiscences about them, if they were not).

The men of 2500 should long to get at the corpses, *but* it's not a question of reviving a handful of various eras. The question is: Will all 370 million be revived? Can we seriously imagine that the people of 2500 will be delighted to make room for 370 million, find them all houses and jobs, give them all job training, and teach them to live in a new society?

Are you uncertain as to what their attitude would be? Well, what would *our* attitude be? Suppose we had 370 million people who had lived anywhere from a generation ago to the time of the generation of Christopher Columbus. And suppose we could revive them and make them part of our society. How would *you* vote?

Do you think we have room for them? Can we find jobs for them? Would the unions object? What about housing? Who would teach them to use internal plumbing and explain to them why their religious views will no longer do?

It could be done, I'm sure. But would you vote to take the trouble, if you could avoid it by just letting them sleep in their cylinders?

Would you be interested in picking and choosing and bringing back a few great men anyway? Suppose you had to pick the great men from the past whom you wanted to bring back to the present.

You could bring back Eisenhower, F. D. Roosevelt, Lincoln, Jefferson, and Washington. Which one would you care to elect President, and how fast would he get used to the modern world so he could even understand what was going on, let alone make decisions? (Let's draw straws to see who explains to Washington that the regulation of stagecoaches is no longer a matter of the first importance.) And what would President Nixon have said when he was President? How anxious would he have been to have Eisenhower back? I mean, really?

Who else would we have back? Shakespeare? To write *The Tragicall Histories of Andrewe Johnson*, Part the First? Moses?

Jesus? Buddha? Shall we vote? Who would you like to have back and why?

Of course, maybe we can't judge the men of 2500 by ourselves. We're mean and selfish, and we wouldn't want to absorb 370 million difficult strangers, but the men of 2500 might be different.

What kind of a world would we have in 2500? Suppose the people of 2500 are living amid the ruins of a civilization killed by overpopulation, pollution, and nuclear war. What could they do with 370 million people?

But no, we can't argue that way. If civilization is ruined, the technology won't reach the point of being able to revive the dead, will it?

If the dead are to be revived at all, then the world of 2500 must be at a considerably higher technological level than our world. That would mean they have survived the present crisis and have established a world government in which a population plateau has been reached, resources are carefully recycled, cheap solar power or fusion power is available in quantity, and pollution is controlled. Total world population may be 1 billion healthy people, well adjusted to their life. Now, that would be just the kind of civilization we would *want* to come back to life for.

But what would the people of such a world do with respect to the frozen dead? They have a carefully controlled population level, a carefully engineered environment kept at a delicate balance. It could be that the last thing in the world they would want would be to bring back 370 million people from past ages, increase their population by a perilous third, and undertake the colossal task of retraining and re-educating.

We could argue that regardless of the difficulties and dangers, they would have to bring back the dead out of simple selfishness. After all, when the men of 2500 die, *they* would want to be brought back to life someday, too. How could they be sure of being taken care of properly if they themselves had set a callous precedent of ignoring the rights of the waiting dead?

Can we really rely on that, though? A society so advanced in medicine that it can bring back the dead and cure them of whatever ails them, could obviously keep those already alive from becoming sick with whatever might kill them. The greater the medical power to restore life, the greater the medical power to stave off death.

In other words, if the world of 2500 could bring us back, they would have immortality for themselves. The immortality might not be absolute. An occasional individual might die of so intricate a disease or so elaborate an accident that even the men of 2500 could not help; but then a relatively short freezing period might fix that up and produce still additional advances capable of handling the unusual event. As time went on, individuals would have to resort to freezing fewer and fewer times for shorter and shorter intervals and, in general, we can say the world of 2500 was immortal enough.

If they have immortality for themselves, they don't have to worry about what is going to happen to them when *they* are frozen, so they don't have the selfish motivation required to revive the dead. And they might prefer to keep their world comfortable and not concern themselves with 370 million dead. What would *you* do if you were they?

On the other hand, the men of 2500 might be humanitarians, and they might find themselves unable to ignore the waiting dead. They couldn't wake them all at one time, but suppose they decided they could handle 10,000 a year—revive them, absorb them, teach them, and make them members of twenty-sixth-century society.

At the rate of 10,000 a year, it would take 37,000 years to complete the job, but what's time to an immortal?

So in the end there are 370 million revivees, most of them seventy years old or more, with heart valves patched up, muscles shrunken, joints rusty. . . .

But no, we mustn't argue like that. If we're going to have lots of faith, we must have *lots* of faith. If we assume that men have attained a level of medical technology capable of returning life to the frozen hulk of a cancer-ravaged body—and curing the cancer, too—then obviously they can cure old age as well. All the bodies brought back will be adjusted to the prime of life; all will be strong, good-looking, and ready to take their place in the new society.

And what about the brains? They would be crammed full of memories, associations, conditionings—everything designed to suit them for the old life. Would it not be hard to uncondition and recondition them, unassociate and reassociate them? Would they ever attain the level of adjustment to the society that the native-born would have attained? Would they not always be second-class citizens, greenhorns, people speaking and thinking with an accent?

These are silly fears. We're speaking of medical supermen, remember? The brains will be dry-cleaned. Nobody today can conceive any possible way in which this can be done, but the people of 2500 will know. They'll take out all the old thinking chains and present the revived individual to the new world with a bright, new, spanking-clean brain, ready for anything.

Ought we go too far in this direction, though? If we take out too much, there will be lost all the old-world memories, and the revived person will be, to himself, a new person with a memory that does not go past the present world he finds himself in. The old person who paid for all that freezing might as well be dead. The body is back, but not the *person.*

We will just have to leave enough basic information so that the revived body can feel a kinship with the body that was once frozen, and then he'll melt into the society of 2500 (or whatever year it will be in which mankind will gain immortality and learn how to revive and cure frozen bodies—the two things go together, remember).

After that, no one will need freezing anymore. It will turn out that the whole episode of freezing will have proved to be a temporary episode in human history. In the passage from mortality to immortality, there will have been an intermediate period of delayed immortality, and some lucky people who chose the pathway of freezing will have jumped the gap of time and joined the fortunate and great last generation of mankind.

And it will be the last generation, for it will be the generation that will live forever and that will therefore have no babies (unless they decide to colonize the Moon and Mars and have enough babies to replace those who travel to other worlds).

Won't it be great? Immortality at last? Life forever? No dear ones dying?

Of course, after you get to be about three hundred years old or so, you may be awfully tired of all your friends and lovers. They will surely have said everything they could possibly have to say about ten times over. You will have, too. In short, a world of immortals would be a world of boredom intensified past endurance.

(I may be pessimistic here, but I'm judging that intensity of boredom from my observation of what happens to the average person on a rainy Sunday afternoon when the TV goes on the blink. The capacity for boredom in three dull hours gives a hint of the colossal height it can reach in hundreds of dull years.)

I've forgotten! The immortals have the powers of unclouding men's minds. Every one of them can go to the mind plants every half century or so and get all the clogged passageways unclogged. Each can start all over again as a bright teen-ager, brainwise, with, of course, all his basics retained so that he knows he's the same person.

Then he can learn again, enjoy again, delight again—and end up after a while as the same bore he was before, because he has the same brain he had before.

And with everyone keeping the same brains (dry-cleaned or not), everyone would retain the capacity for certain thoughts, certain ideas, certain intuitions, and no others. The entire Earth would become a humdrum affair, with all the juice sucked out of it.

I don't see such a world possessing important initiatives. I don't see it populating the universe; it would be too much trouble. It would lack the people, the "new blood," who would have the kind of unexpected daring to make the unexpected jump into the unknown.

In fact, I suspect very strongly that if immortality were possible, no one would take it for long. At varying periods for different individuals, there would be suicide, and this might increase in popularity, to the point where the race would dwindle and shrivel and die, not even with a whimper, but a sigh—and the frozen and revived with them.

Must we say good-bye to immortality, then? Not at all, for we *are* immortal. Not ourselves as individuals, but life is. It has lasted for three billion years now, and since every speck of life comes from a pre-existing speck, there has been a continuous and unbroken thread of life all those billions of years. Who knows how long it will stretch into the future?

This kind of immortality that we have is enterprising and nonboring because it involves a continuous turnover of individuals of all kinds.

You see, we don't have to indulge in vague and rosy dreams of a future in which we can imagine immortals with the capacity of cleaning out old brains and setting up young ones again. We all have that capacity right now. Every time any of us produces a baby, we have produced a young brain.

And more than a young brain—a *new* one. A really new one; one that has never existed before.

It takes two to produce a baby—a mother and a father—and each contributes an equal quantity of genetic material. The baby is a combination of two individuals and therefore contains genetic material that is different in detail from that possessed by any individual who ever existed before.

Every generation we produce quantities of new babies with new brains (billions of them in the present generation), and every generation these are sifted out by the difficulties of life so that only the most suitable or the most resilient or the most versatile or the most something survive to pass on their genetic material in still new combinations to their descendants.

As a result of this shuffling and reshuffling of genes and chromosomes (with a new nonrandom selection made each generation by the forces of the environment), evolution takes place. In the space of a few hundred thousand years, for instance, the brain of the humanlike creatures tripled in size and made us what we are today (such as we are).

If, for the same period of time, there had been individual immortality, there would have been a single generation getting its brains dry-cleaned periodically, there would have been no advance, no change.

So what are we struggling for? As individuals, we have the

right to demand better health, freedom from pain and weakness and depression, active minds, and vigorous bodies into old age.

But have we a right to demand immortality?

No, for that would be a betrayal of the species.

A serene and peaceful death when the time comes is the proper contribution an individual can make to the species that gave him life.

So I have no intention of having myself frozen in the hope of snatching additional life, at the cost of money and energy that can be put to better use, at the cost of intruding upon a future generation that will not want me, at the cost of contributing to the stagnation of mankind.

No! Better, when my time comes, to go with as much of a smile as I can manage and with the hope that others better than myself will take my place and others better than them will take their place and others still better—and so on, forever without end.

Chapter 41 • FOREWORD

I end with these particular last two essays because the life of a doom crier is hard. He is ignored when the doom has not yet arrived, and he is resented when it finally comes. What bearer of ill tidings was ever welcomed?

Naturally, then, why not end with optimistic views? Note, however, that I keep my integrity. The optimistic views I picture in these last essays are clearly stated to be possible *only if the population problem is solved before it destroys us.*

The two articles overlap somewhat, but not completely, and the reader is free to compare and contrast the two and see where my views have been extended, where contracted, where changed.

To guide him there are two chief differences between the two: the matter of when they were written and for whom.

The first of the two essays was written in November 1972, the second in April 1974. Between the two, the energy crisis of the winter of 1973–74 took place, and I think you will see that fact reflected in the second.

Then, too, the first essay was written for *National Wildlife*, with a readership that was not too tuned in to the future and more anxious to remain unassaulted by harsh predictions. The second essay was written for *Galaxy*, a science fiction magazine with a readership well hardened to the difficult, the changed, and bizarre. To those Gentle Readers of critical bent, eager to study the manner in which style is changed to suit the audience— here's your chance.

41 · A.D. *3000*

What do the next thousand years hold for us?

It is impossible to answer with confidence, because some of that answer depends on the unpredictable events that may happen. It is conceivable (if not very likely) that there may be an invasion from outer space. It is also conceivable, for instance, that a new strain of virus might inflict a destructive pandemic on the human race.

Then, too, some of the answer depends on what we ourselves choose to make happen. Mankind might choose to indulge in a thermonuclear war, and obviously the future would then be radically different from that which would arrive in the absence of such a war.

Unpredictables, whether we ourselves are responsible or whether they are imposed on us from outside, are very likely to be catastrophic in nature. If such things happen, we would expect to see a drastic disruption of our society, a breakup of our technology, a great deal of death and destruction. How long it would then take us to pull out of the trough we can't say surely, but even in A.D. 3000 mankind might well be still trying desperately to restore what had been lost.

But let's rule out catastrophes as being by their very nature unpredictable. Let's suppose that life on Earth continues with no surprises, that the course we take is something that can be foreseen because it arises out of what we are doing now. In *that* case, what will life be like in A.D. 3000?

To assume a future of "no surprise" means that life will continue to change in the same way that it is changing now; that the direction in which we are headed does not swerve; that everything, so to speak, will be the same, only more so.

And what direction is that from which we will not swerve?

In the quarter century since World War II and the development of the atomic bomb, there have been three important directions of change.

First, the population of the world has been rising smoothly and, indeed, at greater rate than ever before in history. This rise has been particularly marked in the underdeveloped nations of the world where, for the first time, modern medical techniques—involving the use of insecticides and antibiotics, for instance—has cut the incidence of communicable disease and therefore lowered the deathrate.

Second, there has been the collapse of European domination of the world. Independent black nations have arisen in Africa. Five independent nations now exist where once the British Indian Empire was to be found on the map. China is a Great Power for the first time in modern history, and Japan, despite catastrophic defeat in World War II, is stronger economically than it has ever been before.

This change makes itself felt within the older nations as well. Portions of long-settled nations that feel themselves different from the rest are now vibrant with self-consciousness. The blacks, the Indians, and various ethnic whites are making themselves heard in the United States. The Catholics in Northern Ireland, the French people of Quebec in Canada, the Flemish in Belgium, and the French cantons in Switzerland are all restless. There are religious, language, and tribal frictions in many of the new nations. The world is becoming increasingly prey not to nationalism, which is bad enough, but to what I can only call "localism," which is far worse.

Third, there is a change that is of a sort that has been continuing for two centuries now, but is now proceeding at a remarkably hastened pace. That is the heightened intensification of the technological level the world over. The burning of oil and consumption of metal, the expenditure of the world's resources generally, is proceeding at a rate that no one would have foreseen in 1900, and perhaps not even in 1940.

Suppose, then, that these three changes—increasing population, heightened localism, and intensified technology—continue in the same direction in which they have already been going for thirty years. What will happen?

If technology is to continue to advance, it can do so only at the price of using energy at an ever-increasing rate. If we consider only the energy sources readily available to us today, that means we will burn ever more coal, oil, and gas, and build ever more nuclear power plants.

That presents us with some difficulties. The continued utilization of present-day sources of energy cannot help but change the environment undesirably. The burning of fossil fuels produces air pollution; the use of nuclear power includes the danger of radiation pollution. Furthermore, the advancing level of technology means the ever greater production and use of materials that involve the chemical pollution of soil and water in the process of manufacture, or waste pollution after the materials have been used and thrown away.

Attempts to slow the rate of technological intensification in order to prevent such pollution are hopelessly crippled by the pressure of steady population increase. There are more and more people each year demanding their share of the good things of the world—a demand that can only be alleviated, it would seem, by a continued rise in the level of technology and a continued rise in pollution. Again, the pressure for increased food will place an ever-greater strain on the world's soil, even as it is being destroyed through pollution.

A decision to deal intelligently with *both* technology and population, and to put an end to the era of unlimited growth in both, is hampered by the jostle of new self-consciousness on the part of those who consider themselves (with considerable justice) to have been oppressed and cheated until now. The poorer nations of the world are indignant at the suggestion of a freeze in growth at a time when the advanced nations have a high standard of living and they a low one. That would be, to them, like making injustice eternal.

So if we continue as we have been going, with population increasing, technology intensifying, and localism becoming more important, what can we expect but catastrophe? Pollution will continue to destroy the ecology, and resources will continue to disappear. Mutual friction will continue to absorb human energies and will intensify as wild competition arises in the scramble for what remains of the vanishing good of the world—and that will end only by speeding its destruction.

How much time do we have, then, if we continue as we are going? That is hard to calculate to the minute. If, however, we think of what has happened to our school system in the past thirty years, our judicial system, our political processes, the state of our cities, the safety of our streets, the prevalence of welfare cases, drug addicts, alienated youth—all despite a steady increase in the statistics of prosperity—we can only wonder if we will last another thirty years, let alone a thousand.

With or without a thermonuclear war (and one may be forced out of the hatred and pressures that surround a collapsing society), barbarism looms. Once the delicate balance of our unimaginably complex technological society totters and begins to fall, it will continue to do so at greater and more catastrophic speed, crushing billions to death by famine, by epidemic, and by violence.

What will then be left will represent a favorite plot of science fiction writers—roving bands of primitives lost among the giant wreckage of the past, inventing legends to account for the vast ruins, unable to understand the books they can find or to make use of any but the simplest tools they can salvage—and unable to replace those, once broken, except by further salvage.

Would there not be a recovery eventually? Perhaps, but if the catastrophe is hastened by a thermonuclear war, the level of radiation may, at worst, make human life impossible in much of the world. If the new barbarians find themselves in a world in which the oil is gone and the richest pockets of coal burned and the concentrated lodes of many metals emptied and spread thinly over the face of the Earth, the very basis for a technological society may have been destroyed.

In A.D. 3000, then, people may *still* be trying to raise themselves to the material level we possess today and may not have succeeded. They may *never* succeed.

Am I being too pessimistic?

Of the three great changes, two—increasing population and heightened localism—can surely produce only dangers. The third, however—intensified technology—has a hopeful side. May it not yet bring us advantages undreamed of, advantages that will more than neutralize all the evils we fear? Such an advancing technology may produce unpredictable surprises that will work to our benefit and not to our destruction.

For instance, even if we eliminate unforeseeable surprises, it remains reasonably likely, on the basis of what we *now* know, that, by the turn of the century, we will have reached the possibility of controlled fusion power or learned to make commercial use of solar power, or both. In that case we will then have all the energy mankind can reasonably need for the rest of his existence on Earth. Fusion power will produce no chemical pollution and offer only a minimum chance of radiation danger (far less than does contemporary fission power). Solar power, if use is made only of the sunlight reaching Earth's surface, will produce no pollution of any kind.

What else? Might we not learn to use the ocean as an unlimited source of fresh water, and an endless source of minerals, and a rich source of new food? Might we not treat our wastes in a fusion furnace that will reduce it all to a mixture of indestructible elements that can then be reused, thus ending pollution forever? Might we not solve the puzzle of photosynthesis and learn how to use solar energy directly to produce food out of carbon dioxide, water, and minerals?

If so, does it matter that our population continues to increase? And if we live in a world of plenty in which all can share, won't localism vanish, at least in its dangerous aspects? There might be rivalries still, but surely the kind of mutual hatred that ends in bloodshed would be removed if all sides are sufficiently well fed, well clothed, well housed, and well entertained to make it difficult to entertain the idea of injustice.

That sounds good, but let us consider . . .

At the present moment, the world's population is increasing at a rate that will cause it to double in 35 years. How long can it continue to double every 35 years if it is assumed that technology will double, and even more than double, in inventiveness and ingenuity every 35 years?

Forever?

No, there are limits, even if we allow technology all imaginable advances.

At the present moment, the total weight of humanity on Earth is about 180 million tons. If you double that in 35 years and then double it again in another 35 years and so on, you will find that by A.D. 3530, the total weight of humanity will be equal to that of the whole Earth.

To imagine this is to imagine a completely impossible situation, which means that *no matter what technology does*, mankind cannot continue to increase at the present rate for even as long as 1,500 years. No way!

You might think this is only so because we are making the possibly false assumption that mankind will remain restricted to the planet Earth. You might think that if we imagine technology to advance without limit, mankind will learn to reach other parts of the solar system in any numbers and that it will learn to engineer the other planets so that they will be as inhabitable for mankind as Earth is. Then, surely, we can continue increasing in population at the present rate for a much longer time.

Not so. We can imagine people transferred by the billions to other planets, but we can still calculate that by A.D. 4000, the total weight of humanity, increasing steadily at the present rate, will equal the total weight of the solar system, even including the Sun. And by A.D. 6800 it will equal the total weight of the known universe.

So we've got to stop sometime. Quite apart from any practical consideration, any calculation of resources or energy supply or technological advance, we *will* reach a point at some time in the not very distant future when the increase in human population *must* stop, no matter what technology does.

But suppose that population continues to increase for a while anyway, and that technology keeps up for that while. Suppose it does so for a little less than 400 years, about 40 per cent of the time it will take us to reach the A.D. 3000 mark, which is the goal of this article.

At present, the over-all average density of population on the land areas of the world is about 73 people per square mile. That doesn't sound like much, but of course it is an average. The density in Antarctica is zero, but on Manhattan Island during the working day it is 100,000 per square mile.

Well, by A.D. 2350, at the present rate of increase, the over-all *average* density of human beings on the land surface of the Earth will be 100,000 per square mile. Even if mankind were spread out evenly over the face of the land, the continents and islands of Earth would be one extended Manhattan.

Even if we assume that technology can handle that, it would mean that, at the present rate of increase, less than four centuries (the length of time that separates us from the first settlement at Jamestown) would see the complete disappearance of the wilderness. In fact, greenery—bare, unpaved land—would disappear altogether, except for rooftop gardens.

To feed man's mighty number, no competition could be allowed from other animals, so the Earth would end up with man as the only land animal. If man had to live on naturally grown food, he would have to be a vegetarian and grow food that offered a minimum percentage of waste and indigestible content. That might mean living on yeast and other micro-organisms. Or he might be forced to kill off all animal life at sea and save all the plankton of the ocean for himself.

And where would mankind go from there? Where would he be when we reach A.D. 3000? If the philosophy of growth continues to try to make men expand their numbers and exploit the Earth ever more intensely, we might imagine the colonization of the ocean floor. Even at a drastically slowed rate of increase, the entire surface of the Earth, *land and sea,* could be covered by a Manhattan-density average of 100,000 people per square mile by A.D. 3000.

What next? Can we make our food chemically out of sunlight and simple chemicals? The planetary supply of plants currently uses only one fiftieth the energy of sunlight; can man learn to use all of it and supply himself with 50 times the quantity of food the plant world could possibly supply?

If all the Sun's energy is converted into food, the combined mass of mankind will radiate the Sun's heat over all the Earth. If men are evenly spread out, all of Earth will be equally warm, but any concentration will cause an unbearable heat wave.

And, of course, there will be no room for any other form of life, plant or animal. It will be Earth and its thick covering of mankind, in many layers, all over, and nothing else.

Actually, if we retreat from fantasy, we must agree that at any *reasonable* growth of technology, such a world could not be supported either in A.D. 3000 or at any other time (and even if, in fantasy, it could, no sane human being could want to live in such a world).

Let's put it bluntly, then. Advancing technology is useless if population growth is not stopped; it will only prolong the agony. Nor is it reasonable to hope that the murderous frictions of localism can possibly be soothed and calmed down as long as the pressures of a steadily rising population continue to crush our teeming metropolitan areas and further exploit our deteriorating soil.

If, then, we are going to try to visualize a future without catastrophes, we *must* have a nongrowing population arriving not in centuries, but in decades. By the year 2000, world population should have steadied itself, or the mad slide to catastrophe may well be under way and may by that time prove unstoppable.

By A.D. 2000, though, even if sane population measures are applied as quickly as possible, world population will have reached the 7 billion mark (in the absence of war or famine), and undoubtedly that will be well past the optimum value for the planet's well-being. Considerable damage will have then been done to the Earth, and there will be strong moves not merely to keep the world's population from increasing further, but actually to decrease that population to some more bearable level.

But how can Earth's rate of population increase be reduced to zero and then turned into an actual rate of decrease?

There are, as it happens, only two possible ways. There must be either an increase in the average deathrate of mankind, or a decrease in the average birthrate (or, of course, there may be a combination of both).

If we strive to increase our numbers with total irresponsibility and if technology doesn't keep up (as it surely won't), then the pressures of localism will produce increasing chaos and riot. The deathrate will go up automatically as people die by famine, disease, and violence. But that is what we mean by catastrophe, so that if we are to assume a future without catastrophe, we must assume population control by some means other than a rising deathrate.

This leaves us with a declining birthrate as the only way out. Anyone who doesn't think the birthrate ought to decline, or who thinks it ought to but that there is no practical way of insuring such a decline, automatically must expect inevitable catastrophe.

Not many years ago, I was one of those who expected inevitable catastrophe, but now I have begun to take heart.

Partly through technological advances in the art of contraception, and partly through increasing understanding by the public of the dangers of overpopulation, the birthrate in the United States has dropped to the point where the population growth is small, and it may drop further yet. The birthrate is showing signs of dropping elsewhere, too. Even in China, the very epitome of teeming mankind, there is a drive on for limiting births, and it may be working.

If this trend continues and gathers force, it could be possible to look forward to a stable world in which technology, advancing at a reasonable rate, will have a chance to deal with problems that will then no longer be insuperable. And in that case, mankind will have a chance to reach the year A.D. 3000 with a functioning technological society—and one that is not merely as advanced as ours of today, but even much more advanced.

Now, then, what will such a world of A.D. 3000 be like?

Presumably, it will be a world that will have learned its lesson from the wild chaos and destruction of the twentieth century and the slow recovery of the centuries that followed. It will have settled down to a stable society and will dread above all the results of a population increase. With a total world population of perhaps 1 billion, it will have a deathrate even lower than that of today, but a birthrate equally low.

It will be an elderly population by our standards. The century mark may still represent the extreme of human life, but more people will be approaching it and in greater comfort. Combine this with the low birthrate and there will be a relatively small percentage of children and young people.

The tempo of change may well have slowed as far as social institutions are concerned when man lives in a well-established, mature society, but this ought not necessarily mean that creativity will cease in all aspects of life.

In physical characteristics, of course, mankind will have changed little in a mere thousand years, but mental attitudes are a product of culture. The world will be less youth-centered, less work-centered. With a smaller population, there is bound to be greater room for individualism, more options for each.

With a smaller population and an advanced technology, a much larger percentage of mankind can turn its efforts toward a constructive leisure, toward making something more of life than a drudging attempt to earn a bare living or a manic effort to achieve a bare amusement. The percentages of artists, writers, craftsmen, athletes, scholars, and showpeople to the general population will surely be much higher than today.

There will, in short, be more options of greater variety for individuals, and this is as good a definition as any I know for "progress." Simply because the work of the world will be so largely given over to machinery, the individual can spend his time designing his time to suit himself. If he wants to gather a technical education and devote himself to what we would call "work" today, he can—but he wouldn't have to. There will be enough, surely, who will choose to do the work of the world as *their* thing, to keep the world running.

Women will have achieved full social and economic equality with men automatically, once the population becomes stable. Since women would not be expected to have many babies through an extended lifespan, and since even those may be brought to term outside the body and then brought up in communities, rather than in private, what can we expect women to do with themselves?

If they are not allowed to take their place beside men in as equal a fashion as possible, they will be driven by boredom to attempt to have babies, and this is what the society of A.D. 3000 will be most concerned to prevent. Women will therefore be encouraged to occupy themselves in all activities—whether it be the necessary work of the world, or a personal hobby, or sheer amusement—in the same way that men will. And given such equality of opportunity, there is every reason to suppose that women will amply demonstrate their intellectual and creative equality to men.

At least one technological advance that will be basic to the world of A.D. 3000 will be the communications satellite, which, after all, already exists.

Long before A.D. 3000, we have every right to expect that communication between any two points on Earth will be as simple as shouting across the street. By using laser beams to streak to the communications satellites and back, millions of channels (radio, television, and telephone) can be crowded into use. Every person on Earth can have his own wavelength, his own private circuit. Information transfer will have made it possible for each person to tap the resources of a world library and reproduce the pages of any book he wishes, or any newspaper. All cultural events taking place anywhere on Earth can be readily available, either at the times they happen or (by cassette) at any time afterward, at any other point on Earth.

This means that the social pressures to gather in large concentrations of population will decrease. Since people can gather in conference by means of closed-circuit television without having to budge from home; since factories and offices will be increasingly automated and robots may well be doing the nondemanding muscle- and brainwork of the world, the economic pressures will decrease also. Long before A.D. 3000, the city may have withered as a human institution, and people will be living in relatively small and scattered communities.

Population will be better distributed. Advanced technology, including cheap and abundant energy, and advanced control over weather, will make few regions of the world uninhabitable. It will be as easy to procure the necessities of life in a valley in the Andes as on the Hudson River, and the intellectual and cultural aspects of life will be easily available in both places.

Nations will exist as marks on a map but there will, effectively, be a world government, and the population will have spread out and become too mixed to allow much in the way of nationalistic folly. The more even spreading will include a leveling of economic differences as well, so that pockets of poverty and other pockets of wealth will not exist. What differences exist will be differences of way of life by choice. People will be different, but only as they choose to be and not through externally imposed force of circumstance. Thus, they may speak different languages if they choose, but they will all also be able to speak what we might call "Standard Planetary," which may be more like English than like any other language.

Is all this merely a visionary ideal? Perhaps not. If the population level is controlled and if technology is given a chance, the world will surely grow increasingly computerized. Computers, making decisions on the bases of accurate inputs and rational programming, will never be so emotionally unstable as to suggest measures that will increase the possibility of war, or be so unrealistic as to imagine that a stable society can be based on mass injustice.

Guided by computer reason, rather than by human emotion, we may end in A.D. 3000 in a world in which not only will women be granted equality of opportunity, but also *all* individuals will. With a smaller population on Earth, greater decentralization, less crowding, more room, instant communication, rational computerization, and plenty for all, localism would lose its bite.

People may still be different in taste and thought, and one individual may disapprove of or even detest another, but mere differences will not lead to psychotic action. Indeed, in a society in which genetic engineering will be greatly advanced, differences among individuals will be considered the greatest single resource of the human race, for those differences will represent a large and varied gene pool—an incalculable advantage to the species.

In A.D. 3000 less of Earth's surface need be given over to farmland. A total population of 1 billion will require less food than we do today, and the farms that do exist will be, again, automated and require little in the way of human supervision. By A.D. 3000, weather control will be practical, and that, plus a far better understanding of soil chemistry and physics, may make it possible to produce the necessary food in better quality on perhaps one tenth the land given over to the purpose today. This will be all the more true since the sea will be efficiently farmed and there will be a significant amount of synthetic food available.

Very likely the larger and more dramatic species of land life, especially the larger predators, will be gone. Man will be forced to substitute and serve the purpose for a healthy ecology. Controlled hunting will keep down the population of the larger wild herbivores that remain and make game of various sorts a more common item in the dietary.

The result of all this will be that a much larger portion of the world may be "wilderness" than is true today. I put the word in quotation marks because it probably will not be much like what we call wilderness today. Those species that man has wiped out and is wiping out cannot be restored; and some of the damage man has already done and will yet do to the environment may very likely not have been healed even by A.D. 3000. Then, too, although mankind will be relatively few in number, it will scatter, and there will be few areas where human beings will not live and that they will not attempt to adjust more closely to their comfort.

The world of A.D. 3000 may therefore resemble one huge national park or, rather, planetary park, in which plants and animals may flourish in perhaps greater numbers (if fewer varieties) than today, but which will all be closely supervised.

By A.D. 3000, there may, in fact, be a tendency to choose to build communities underground. That would involve certain advantages. There would be escape from temperature extremes and from storms. Transportation would be easier, and time adjustments would be more rational in the absence of an alternating day and night of varying lengths through the years.

Nor would this be a retreat from the open air; rather the reverse. In any underground community, the wilderness would be available at the end of a short elevator rise. It would be far closer to the average man than is true for one in the center of any of today's cities. And that wilderness would not be distorted and destroyed by man's immediate presence.

Abundant energy of a nonpolluting kind would keep the wilderness clean and healthy, for one thing. The ability to make adjustments in the environment would also mean that man would deliberately try to produce as much variety as possible in the wilderness. There will be wooded areas, wet areas, and whatever else might be required to insure a balanced ecology and one that would give as much pleasure as possible to man.

Does such a life, presented here only dimly, seem too controlled, too tame, too dead? Does it seem unpleasantly bland to those of us who are accustomed to change and uncertainty? Do we prefer the heady risks of disaster, the casual flirtation with catastrophe, to a dull "domestication?"

If so, remember that I am so far speaking of Earth only. After all, by A.D. 3000, Earth will surely long since have ceased to be the only home of the human race, even if technology advances at only a moderate and realistic pace.

The Moon will be an old and well-established colony of Earth. By A.D. 3000, it will have been virtually fully exploited and will support a large population, which will lead as advanced and sophisticated—and unadventurous, perhaps—a life as men on Earth.

There will, however, be important differences on the Moon as compared with Earth. For one thing, the Moon colonists would *always* have lived underground, *always* have experienced a thoroughly engineered environment, *never* have known an open wilderness.

Their world will be closer in principle to the microworld of a spaceship, and it would, therefore, be easier for them to adapt to long spaceflights. Furthermore, the surface gravity on the Moon is only one sixth that on Earth, and the escape velocity from the Moon is only one fifth that from the Earth. That means that a rocket liftoff from the Moon's surface is much easier and requires much less in the way of thrust and acceleration than on Earth. Nor is there an atmosphere on the Moon to offer resistance and produce dangerously high temperatures.

It would seem likely, then, that once the Moon colony is well established, it would prove to be the Moon colonists who will further the exploration of the solar system. By A.D. 3000, they will have established a flourishing colony on Mars and will have outposts on the larger asteroids and perhaps even on some of the satellites of Jupiter.

Indeed, we can say that by A.D. 3000, the solar system will be reasonably well exploited.

And if the core world of Earth, together with the old colony on the Moon, are staid and settled and have grown to drowsy maturity, then vigor, adventure, and danger will yet be found on the further worlds and in the thoroughly artificial space stations that may fill space from the uncomfortable neighborhood of Venus to the frigid distances of Saturn.

By A.D. 3000, in fact (if not before), mankind will be ready (with the aid of what technological advances I cannot clearly foresee) to make the long voyage to the other stars. Having burst out of the prison of Earth, mankind will next burst out of the larger prison of the solar system as well.

And out among the unimaginable reaches of the interstellar spaceways, up the long climb from star to star, mankind can yet grow and expand and find all the joys of danger and insecurity for as far into the future as my eye can see.

42 · Is There Hope for the Future?

At any time, and under any conditions, it is possible to consider the future either pessimistically or optimistically.

It is, after all, never possible to predict the future sharply; one can only make an estimate, and that estimate will cover a range of possibilities. The range will be broader the farther into the future we look, and if we look far enough, the range becomes so broad that our predictions have no constraint worth mentioning other than the laws of nature. The range will also be broader as we deal with more and more poorly understood phenomena (with human psychology, for instance, rather than with atomic physics), until it becomes too broad to make prediction useful.

If, however, we restrict ourselves to the moderately close future and to moderately well-understood phenomena, we end up with a range of possibilities that is not prohibitively broad. We are then at liberty to suppose ourselves anywhere within the range, and it is possible to end up with a pessimistic prediction if we choose to use one extreme of the range, and with an optimistic one if we choose to use the other.

It is, for instance, particularly easy to be pessimistic about the future right now. We need merely assume that population will continue going up; that national rivalries will continue to place the well-being of Group X ahead of the welfare of the world; that racist and sexist prejudice will continue to generate hatred and alienation; that personal and economic greed will continue to ruin the Earth for short-term private profit—in short, we need merely assume that things will go on, exactly as they have been, for another thirty years, and we can confidently predict the destruction of our technological civilization.

I suspect that the chances are better than 50 per cent that this will happen; how much better, I am not certain.

But things don't have to go on as they are. Things *do* change, and surprisingly rapidly, too.

Place yourself in 1954, for instance. It is the height of the complacent Eisenhower era; the depth of the Dulles cold war. The United States was then at its most self-confident and sanctimonious point in history.

Would you have imagined, then, that over the next two decades, contraception would have become socially acceptable; that a birth control pill would lead to a sexual revolution; that abortion would become legal in many places; that "cold war" would become a dirty phrase, and that the ineffable Richard Nixon who had, in earlier years, specialized in flag, mom, and apple pie rhetoric would, as President, lead the way to closer friendship with the Soviet Union and with what he then began to call the People's Republic of China?

I tell you that in 1954, it was a lot easier to predict, and to have it be believed, that men would stand on the Moon in fifteen years, than that any of the situations listed in the previous paragraph would come to pass.

Why did all those things come true? No mystery at all. The steady increase in population and the steady decline in resources have faced mankind with a choice of (1) destruction or (2) population control and world government.

The changes that have taken place in the past twenty years have been in the direction of population control and world government and were more or less inevitable if one had been willing to look the future in the face.

The changes so far have been comparatively small and tentative, and are far from sufficient to do the job of preventing disaster altogether. I think it is safe to suppose, however, that mankind will continue to move in the direction of population control and world government as, each year, the scope of disaster and the speed with which it is coming impress mankind generally with a greater and greater horror.

The question is not whether mankind will move in this direction, or not—it will!—but whether it will move in this direction rapidly enough. Again, my own feeling is that the probability of rapid-enough motion is less than 50 per cent; how much less, I am not certain.

But the motion may be faster than I fear it will be. In this respect, the energy crisis of the winter of 1973–74 performed a great service. The crisis was, to a great extent, made inevitable by the folly of American foreign policy since World War II (see my article "The Double-Ended Candle" in my book *Of Matters Great and Small*, published by Doubleday in 1975) and was exacerbated by the greed of oil companies, but underlying everything is the fact that there is only a limited amount of oil in Earth's oil wells and that that limited amount is vanishing with frightening rapidity.

Nothing has done as much as the energy crisis to convince Americans that the economic interdependence of the world is real and that it includes the United States. Nothing has done as much to convince Americans that our standard of living, so much higher than that of the rest of the world, is at the mercy of the rest of the world. In short, nothing has done as much to convince Americans of the *vulnerability* of the United States. That goes a long way toward making world government seem less of a dirty phrase, since it is now obvious that we may have as much to gain from a globally organized economy as to give.

To be sure, at the time I write, the Arab nations have lifted their boycott and Americans are doing their best to convince themselves that, after a three-month nightmare, everything is exactly as it once was—but that's self-delusion. Oil prices have gone up, inflation has moved along faster, and the oil in the ground is still disappearing. The crisis, believe me, is still with us and won't go away, and a certain amount of American innocence will never return.

Let us imagine, then, that the Earth continues to move in the direction of population control and world government and does so quickly enough to avert a major catastrophe, but suffers, at most, only a mild catastrophe. (After a recent talk I gave at the University of Pittsburgh, I was asked what I meant by "a mild catastrophe," and I replied, "One from which civilization can recover.")

This supposition of fast-enough movement may be low probability, but perhaps it is not zero probability. Perhaps, under the lash of the gathering horror, we will be forced, kicking and screaming, into survival.

In that case, here would be the situation as Earth enters the twenty-first century:

1. World population will stand at 7 billion, but all over the world, heroic and successful measures will be holding the line, and every effort will be made to lower the birthrate to the point where the population will decline toward an ultimate goal of perhaps no more than 1 billion.

2. There will be dreadful shortages of food and raw materials generally, but heroic and successful measures toward the proper distribution of what exists and toward efficient methods of recycling will minimize the more disastrous effects of the shortages.

3. There will still be political units of the type with which we are familiar, but few decisions of any importance will be reached except at international conferences. It will furthermore be clear that no nation can afford to take unilateral action against the will of the others.

If all this is so, we can work out as inevitable corollary (or inevitable, at least, as long as mankind chooses not to choose destruction) a number of utopian consequences. For instance:

1. *The end of sexism.*

Womankind's subjection has been the natural consequence of her role as baby machine. In a world of high infant mortality and low life expectancy, the need was for many babies. It takes many babies to have even a few survive, and in an agricultural economy many children mean many hands to help with the work. Children are also needed to help support aged parents in a society that would otherwise let them die. (That is the significance of the biblical "Honor thy mother and thy father." It doesn't refer to standing up when they come into the room. It means *supporting* them.)

In the twenty-first century, with a very low birthrate, with childlessness common, and with those children who are born very much more the responsibility of society in general than they are now, women's role as a baby machine will have largely disappeared.

In that case, what else will a woman have to do? Do you suppose that she can still be relegated to social and economic inferiority; made to accept the situation that household tasks are peculiarly hers; that passivity is her role in sex, in business, in government; that her highest function is to support her man in a self-effacing manner, and that she must place her physiological wares (but *never* her intellectual wares) constantly on view to catch him in the first place and reflect favorably on him in the second?

If this were indeed to be the situation, women would be condemned to lives so empty that child-bearing and child-rearing would be all that could fill them. There would then be an enormous tendency to strive for children under any conditions.

To keep the birthrate successfully low, women must be beguiled into other activities, and what method would be so natural and so effective as to declare them people and to allow them to enter *all* facets of human endeavor on an equal basis with men?

2. *The end of racism.*

Racism has existed as long as mankind has, because any slight difference marks one as outside the tribe and, therefore, as someone to be mocked, if mockery is safe, or feared, if it is not. Introduce a new child into a group of children, and have him wear clothes a trifle different in style, or speak in a slightly different accent, and watch him become marked for scapegoating at once.

It does not matter that the clothes may differ in being cleaner, or the accent in being more precise; the result is the same. The key word is not "better," nor is it "worse"; it is merely "different."

And, of course, in the thought processes of the bigot, "different," whatever its nature, *becomes* "worse."

That is why I am not impressed by the attempts of men like Shockley to argue that blacks are less intelligent than whites; that it is the natural inferiority of blacks that has caused them to be discriminated against, and (by obvious implication) that it is for that reason that they should continue to be discriminated against.

In the first place, I don't accept Shockley's arguments on intelligence. I do not believe that intelligence can as yet be measured, or even *defined*, with sufficient precision as to make it possible to divide humanity into large groups of greater or lesser intelligence, with the difference just happening to coincide with something as irrelevant to intelligence as skin color.

Nevertheless, *if* intelligence *could* be defined and measured, and if it turned out that blacks *were* inferior to whites in intelligence, that would still be totally irrelevant to the matter of the continued mistreatment of blacks. It is the difference in appearance that triggers the bigotry, and it would be no less if the blacks were *more* intelligent than the whites.

As a matter of fact, I know of minority groups which, in the stereotypical minds of the bigots, are stigmatized as being *too* intelligent. They are "cunning," "shrewd," "sly," and, although in a small minority, are continually on the point of "taking over the country" if they have not indeed already done so. And how does Shockley explain that?

But observe how matters will change in the twenty-first century; not out of the increase of goodness and love in the human heart (alas!), but out of the pressing necessity for survival.

If population is to be stabilized and even forced into a period of slow and humane reduction, it can only be accomplished by convincing humanity that this reduction is not an excuse to wipe out some groups and perpetuate others. Birth control can easily be used for this purpose, or be suspected of being used for this purpose.

In order for population control to work at all, and for mankind to avoid catastrophe, then, all people (or at the very least, *enough* people) must be convinced that all groups will be respected equally. While open bigotry exists, how can people be convinced of this? Mankind will simply have to school itself to assume a virtue if it has it not, and pretend to love neighbors and fellow men even when it does not. And if the assumption is made long enough and the pretense is kept up steadily enough, the fact that it is merely assumption and pretense may eventually be forgotten.

Of course, you might imagine that we needn't *persuade* inferior people to cut down on their children. Why don't we just wipe out all those high-breeding, low-standard people and control the population even more efficiently? That might sound nice to you if you're sure that nobody with a plane and a bomb is going around considering *you* inferior, but let's suppose you are on the right side of the gun.

It would still not be the right side, for the policy of wiping out the unworthy would not be merely a matter of powerful countries wiping out weak ones. *Within* every country, if bigotry rules, there are racial and economic groups that would seem, to bigots, to be breeding too fast and to be best controlled by death. The confusion and chaos that the rule of death will then bring about will surely dissolve our all-too-rickety technological structure and bring it down upon our heads, even if we happen to be the ones holding all the guns.

No! If the twenty-first century is to work at all, it will have to be working without racism.

There will be factors that should make this easier than we now think possible. If technological civilization survives into the twenty-first century, it is quite obvious that computerization and automation of society will continue to advance, and such advance will militate against racism.

Increasingly, we will be developing a society in which unskilled and semiskilled manual and mental labor will be done by machines, and there won't be the economic pressure to maintain a large supply of people under conditions of oppression and of carefully inculcated acceptance of inferiority, in order that these might be content to perform these unskilled and semiskilled tasks at low pay.

(Naturally, the disappearance of such work will make it all the more sensible to reduce the population, since it will take fewer people to run the world.)

Then, too, advances in communication—the use of satellites bound to each other and to Earth's surface by laser beams capable of carrying millions of communication channels—will knit the entire globe into a small community ("global village" is the term most frequently used).

While efficient communication is no guarantee of brotherly love, it does make it a little easier to get along with someone you dislike if you can at least talk to him.

The fact that in the twenty-first century it will be far easier for all people to have access to education and to the general store of information amassed by the species will wipe out some of the more obvious and fallacious "intellectual" differences.

In a global village there will also be an increasing push toward a common language. I don't mean necessarily an *exclusive* common language, with all the rich differences in language and culture that now bless our planet wiped out. Let each group have their own language and ways, but let each group also know some language with which they can reach all other groups.

(I personally favor English as the common language, because of its great vocabulary and because of its already unparalleledly widespread use—and also because I am a linguistic chauvinist pig.)

The smallness of the world, the ease of communication, the equalizing of opportunity, and the common language will all act to depress the sense of difference and will therefore tend to defuse the push toward bigotry.

Even the mere fact of a decreasing population in a century of continuing scientific advance will make bigotry increasingly unpopular. The gradual increase in the understanding of genetics will make it clear that, from the standpoint of species survival, the greatest asset we can have is genetic diversity.

There are species that are so perfectly adapted to a particular environment that they survive virtually unchanged for millions and millions of years. Such perfect adaptation achieves relative genetic uniformity and makes those species sharply limited in range and at the total mercy of the environment. Let the favored environment disappear, and the species lacks the genetic equipment to survive.

The genetic diversity of a generalized species makes it possible for that species to adapt this way or that and to survive in one form or another long after the living fossils have met their doom.

As the human population declines, then, there will be considerable concern lest too many genes vanish. People generally will then hail diversity and be glad that other people exist who are different from themselves in appearance and abilities as living proof that the human gene pool is still healthily broad.

3. *The end of war.*

Actually, we have already reached the end of war, as long as national leaders are guided in their decisions by sanity. (That they will be so guided is not a foregone conclusion of course.)

A nuclear war between the United States and the Soviet Union is clearly mutual suicide as far as the two nations are concerned. What's more, it would probably destroy our technological civilization generally and, through the radiation it will produce, will seriously compromise the viability of the planet as a whole.

This almost everybody recognizes, so the question is whether a *non*nuclear war is really possible, and the answer is "No!" The trouble is that the advance of technology has made war into such a high-energy game played with such high-sophistication pieces that no one can afford to play the game anymore.

War is fought with a nation's surplus energy and resources, under the best of conditions; or else a nation can fight a *short* war even without surplus energy and resources in the hope of seizing an enemy's energy and resources and continuing the fight with those. Where *no* nation has surplus energy and resources large enough to support the current technology of war, the whole process becomes purposeless and a mere excercise in suicide, one somewhat slower than that of the nuclear war.

The most recent war that managed to last for years, and that reached a clear-cut decision without too badly damaging the victors, was, of course, World War II. Since World War II (thirty years now!) there have been two wars that involved at least one great power directly and that lasted for years—the Korean War and the Vietnam War.

Both of these wars ended exactly where they began. The United States had to end each war by dealing with an enemy whose territorial extent, military strength, and political nature had not been changed by the American effort. All we could claim was that the other side hadn't actually won. In each case we could have wiped out the enemy if we had exerted our maximum strength, but in each case we did not, on the whole, dare to.

All other wars fought on Earth since 1945 have been small-scale, or very short, or both. And in every case, none of them could have progressed at all without the support given to one side or another by one of the great powers.

Right now, it takes all the United States can spare to support a military force under *peacetime* conditions, and no other nation is any better off. And as energy supplies and material resources decrease, it will become more and more difficult to afford all those uniforms and all that gold braid.

In the twenty-first century, the nations of the world will be forced into international co-operation as the only way of tackling and defeating the problems besetting us, and armies will be expensive anachronisms anyway, except, perhaps, as organized police and labor forces.

So war will vanish not because of a growth of goodness in the human heart or understanding in the human mind (would that those were so!) but only because war has *already* priced itself out of existence, except as a form of world suicide.

4. *The extension of the lifespan.*

If our technological civilization survives into the twenty-first century, that will mean that medical science will continue to advance.

Increasingly, degenerative and metabolic diseases will be successfully treated. Arthritis, cancer, and circulatory disorders may all join the various infectious diseases as merely minor dangers.

This means that more and more people will reach the age of seventy before dying (already in such places as Scandinavia, half the men and slightly more than half the women do so). The twenty-first-century population will then consist of a greater percentage of old people than now exists in the population and (thanks to birth control) a considerably smaller percentage of young people.

Gerontology—the medical study of the phenomena of old age—

will therefore become the most important medical specialty because of a plethora of subjects and a falloff in the importance of other specialties.

Until today, all that medical advance has done is to make it possible for more men and women to grow old. This is not to be sneered at, of course, and I am personally delighted with even such limited progress, since it won't be long till I'll be passing out of late youth and into very early middle age myself.

Still, once a person reaches seventy, he or she is old—as old today as he or she would have been if he or she had performed the much more difficult task of running the gantlet of disease and misery and had reached the age of seventy in Homer's time.

Old age is sometimes said to be just one more disease, but if so, it is disease different from all others, since it alone seems to be inevitable and inescapable. There is logical reason to suppose that old age is built into the genes. Cells from human embryonic tissue, though given an idyllic, protected environment, and supplied with ample nourishment, divide more and more slowly as time goes on, and after some fifty divisions divide no more.

The cells run down; their divisions eventually stop; they die and are not replaced; and the whole intricate machinery of the body falters, then grinds to a halt. The running down may be through the accumulation of errors, as genes replicate time after time; or through the slow accumulation of waste products; or through the slow deterioration of protein molecules.

Whatever it is, it seems to be programmed into the organisms from the start—and we can see why this should be so.

After all, each new child is born with a brand-new gene combination to be tested for its survival value. Each new child is a new throw of the evolutionary dice, a new turn of the wheel. In order to assure the necessary sorting and resorting of the new, so that the species can be always adjusted to better fit an old environment or to come to fit a new one, the old must be taken off the stage. The existence of death by old age, which makes certain this removal even when all other causes of death fail, encourages and speeds evolution. It hastens and strengthens the development of the species at the cost of the individual.

But whatever the cause of old age and however programmed

that cause might be, could it not be reversed, as biologists learn more and more about the intimate details of cellular biophysics and biochemistry, and learn also how to manipulate those details?

Might old age be prevented for a time, or reversed to a degree, and might not people live for two centuries rather than one and remain young through most of the doubled lifespan? Might they not live longer still? Might they not be potentially immortal —or at least have the option of living until they voluntarily choose to die?

Perhaps! It may be that the twenty-first century, while it sees the population decreasing steadily, will also see the individual lifespan increasing steadily (and therefore making necessary a still further drop in the birthrate).

But in that case, will not the extended lifespan and the ever slower addition of new babies to the species slow human evolution and endanger human survival in the long run?

Yet who is to say that evolution must proceed only by that mechanism that has, in fact, been used through all the billions of years of life on Earth? So far, evolution has been carried on by random gene-combinations and recombinations; by random gene-mutations and new combinations; and by an endless epidemic of random death to make sure of an endless turnover of generations with their new gene-combinations.

Now, after 3 billion years, we have on Earth a species that is, for the first time, potentially capable of directing its own evolution in a nonrandom manner.

Perhaps the twenty-first century will see the beginning of something the world has not yet seen, something radically different: a species more stable than any other that has ever existed; one with individuals that endure far longer and remain far less affected by the passing of the years; one that accumulates wisdom and experience in each individual to an enormous extent; and one that guides its own evolutionary destiny across the very slow heartbeat of generations through thoughtful genetic engineering rather than by random death.

5. *The expansion of man's range.*

If we imagine the triumph of genetic engineering, we can, if

we choose to adopt the pessimistic view, picture mankind as consisting of a limited number of very old, very tired individuals, who
for centuries have not had a new thought. You might see mankind
turning from physiological death merely to find a new and infinitely more horrible intellectual death.

Even if we discounted genetic engineering and immortality,
and supposed that death will always hold its sway over mankind,
we might still picture the twenty-first century as the century of the
middle-aged, since there are bound to be more old people and
fewer young people then. Might not the generally older population be stodgier, more conservative, more unoriginal, more unin-
novative, than we are today?

We might even argue that in the twenty-first century, mankind
would have learned the lesson of the twentieth century (or civilization will not have survived). In the twenty-first century, people
would know that indiscriminate growth was no longer possible.
They would know that they could not consume and pollute at
will. They would know that everything would have to be recycled
as far as possible and that every new advance, every change,
would have to be closely examined for side effects.

Conservatism will, of necessity, be built into twenty-first-century society, and that great and heroic dash into the unknown
will be forever gone. What will be left will be the life of the
sloth, which, hanging suspended, moves each limb slowly forward
and tests the branch carefully before gradually shifting its weight.

Is this what we must look forward to?

We can, of course, argue the point. Are old people really stodgier and more conservative than young people? In societies in
which the proper attention is paid to the old, and which are not as
youth-worshiping as our own today is, might it not turn out that
the old are as innovative as the young?

Suppose, though, that we don't argue the point. Suppose that
we *are* threatened with an innate conservatism and the death of
daring. Is there any way that these can be fought?

What we need is a horizon to be passed, a limit to be penetrated. Of course, there will always be horizons and limits in the
intellectual world—and the great battle against the unknown will

never be over. But this is an ethereal battle and one that may not catch at the imagination of humanity as a whole.

What we need is something physical and visible, and surely that we have. When the last horizon on Earth has contracted to zero and the last limit has vanished, there remains an unimaginably vast universe beyond the Earth.

In the twenty-first century, space exploration and space colonization will become not merely matters of scientific curiosity but will be things necessary to keep alive that vital spark of daring in mankind. And in adopting an exercise to insure the survival of the spirit of humanity, we will also gain in very important ways.

On the Moon, a colony can take advantage of the Moon's environment—its airlessness, its extremes of temperature, its hard radiation—to gain knowledge and to develop industrial finesse that would be difficult or impossible to accomplish on Earth.

In addition, a lunar colony, to survive, would have to do so in an environment even more restrictive than that of Earth, and it will serve as an example to be followed. The Moon could easily be the school of Earth.

Then, too, it may be only by way of a Moon colony that mankind can explore the rest of the universe.

The Moon is easy to reach—it is only three days away even by the primitive space technology of today. To reach any sizable body of the solar system beyond the Moon will, however, take anywhere from months to decades; and to reach even the nearer stars will take from decades to centuries. To imagine Earthmen forsaking Earth for years or lifetimes in a constricted spaceship is to imagine too much, perhaps.

To be sure, the Earth is itself a spaceship, but it is an atypical one. It is the kind of spaceship in which the life-support system, and the crew, cling to the outside of the hull, and have grown so used to this that life within the hull is difficult to adjust to.

On the other hand, a Moon colony can only exist in caverns beneath the surface, and that would be in a typical spaceship environment. For a group of Moon colonists to get into a spaceship and venture farther out into space for years at a time would be far easier, psychologically, than for Earthmen to do so. To the

Moon colonist, the spaceship would be much more nearly like home.

And if the time comes when large ships are built that are capable of supporting an ecologically independent human society over the generations, then surely it will not be Earthmen but the Moon colonists—or their descendants, the people of the hollowed-out asteroids—who will serve as the crew.

In fact, we might imagine the asteroids themselves, after having been inhabited for a greater or lesser time, turned into spaceships, driven out of their orbits by some advanced spacedrive, and launched beyond the solar system and into the depths of space. In that case, there would be no psychological difficulty worth mentioning. The crew would be staying at home.

So however stodgy Earth may get (and I insist that it may *not* get stodgy), there will always be the escape value of space exploration, and the twenty-first century may witness the beginning of the expansion of mankind's range—an expansion without limit.

People from Earth may sometimes qualify to immigrate to the Moon; people from the Moon may sometimes qualify to immigrate to one or another of the asteroids; people from the asteroids may sometimes choose to launch themselves into interstellar space.

The net result will be that the Galaxy, and, indeed, all the galaxies, will be opened, in the long run, to human beings and to their descendants (proliferating into many parahuman species). Out in space, humanity in all its varieties may meet and mingle with nonhuman intelligences on their own level, so that we will no longer be alone.

What's more, if tachyons do exist and if we can bend them to our will—or in other ways get around the speed-of-light limit— we may even end with the kind of galactic empire dreamed of by myself (if mankind is the only intelligent species in the Galaxy) or by E. E. Smith (if it is not).

Let's summarize, then. The immediate future looks dark, and civilization may not survive the crisis that is upon us.

If, however, we can shift quickly enough in the direction of population control and world government and can hang on for thirty years, the long-range future—within the later lifetime of the young people alive today—can be made incredibly bright.

We will then have a twenty-first century that will be the dream of an older generation of science fiction writers (writing prior to the current fashion of darkness and doom) come true. Imagine a world in which the scourge of war is eliminated and the horrors of sexism and racism are wiped out, in which lives are expanded and enriched, and in which all of space is opened to us.

If only we can get through *this* crisis. . . .